SpringerBriefs in Geography

For further volumes:
http://www.springer.com/series/10050

Carlo Rega
Editor

Landscape Planning and Rural Development

Key Issues and Options Towards Integration

Springer

Editor
Carlo Rega
Interuniversity Department of Regional and
Urban Studies and Planning (DIST)
Politecnico di Torino
Torino
Italy

ISSN 2211-4165 ISSN 2211-4173 (electronic)
ISBN 978-3-319-05758-3 ISBN 978-3-319-05759-0 (eBook)
DOI 10.1007/978-3-319-05759-0
Springer Cham Heidelberg New York Dordrecht London

Library of Congress Control Number: 2014938989

Printed on acid-free paper

Springer is part of Springer Science+Business Media (www.springer.com)

To Silvia, Francesco and Anna
For your love and patience

There's never been any shortage of landscape in the world. Whatever else may be lacking, that's one thing that has never been in short supply, indeed it's sheer abundance can only be explained by some tireless miracle, because the landscape clearly pre-dates man, and despite it's long, long existence, it has still not yet expired. That's probably because it's constantly changing: a certain times of the year, the land is green, at others, yellow or brown or black. And in certain places it is red, the color of clay or spilled blood. This, however, depends on what has been planted, or what has not yet been planted, or what has sprung up unaided and died simply because it reached its natural end. [...]. But who are these other people, small and desperate, who came with the land, although their names do not appear in the deeds, dead soul perhaps, or are they still alive? God's wisdom, beloved children, is infinite: there is the latifundio and those who will work it, go forth and multiply. Go forth and multiply me, says the latifundio. But there is another way to speak of all of this.

José Saramago "Raised from the Ground"
(The Random House Group, 2012)

Contents

Contributors

Alessandro Bonifazi Independent Researcher, Bari, Italy

Marta Bottero Interuniversity Department of Regional and Urban Studies and Planning (DIST), Politecnico di Torino, Torino, Italy

Chiara Bragagnolo Institute of Biological and Health Sciences, Federal University of Alagoas, Maceió, AL, Brazil

Claudia Cassatella Interuniversity Department of Regional and Urban Studies and Planning (DIST), Politecnico di Torino, Torino, Italy

Rinaldo Grittani Agricultural and Environmental Science Department, University of Bari "Aldo Moro", Bari, Italy

Carlo Rega Interuniversity Department of Regional and Urban Studies and Planning (DIST), Politecnico di Torino, Torino, Italy; Currently at: European Commission, Joint Research Centre, Institute for Environment and Sustainability, Ispra, VA, Italy

Chiara Rizzi Department of Civil, Environmental and Mechanical Engineering (DICAM), University of Trento, Trento, Italy

Bianca Maria Seardo Interuniversity Department of Regional and Urban Studies and Planning (DIST), Politecnico di Torino, Torino, Italy

Stefania Staniscia Department of Civil, Environmental and Mechanical Engineering (DICAM), University of Trento, Trento, Italy

Andrea Tassinari Theorema S.A.S., Venice, Italy

Abbreviations

AES	Agri-Environmental Schemes
CA	Conjoint Analysis
CAP	Common Agricultural Policy
DSM	Digital Surface Model
DTM	Digital Terrain Model
EARDF	European Agricultural Fund for Rural Development
EC	European Commission
ECA	European Court of Auditors
EIA	Environmental Impact Assessment
ELC	European Landscape Convention
ERDF	European Regional Development Fund
EU	European Union
GAEC	Good Agricultural and Environmental Conditions
GIS	Geographic Information System
LChA	Landscape Character Assessment
NUTS	Nomenclature of Units for Territorial Statistics
PTP	Provincial Territorial Plan
RDP	Rural Development Programme
RLP	Regional Landscape Plan
RMA	Rural Mountainside Areas
RTP	Regional Territorial Plan
SEA	Strategic Environmental Assessment
TEV	Total Economic Value
VS	Visual Sensitivity
WTA	Willingness to Accept
WTP	Willingness to Pay

Chapter 1
Introduction: Rural Development and Landscape Planning—Key Concepts and Issues at Stake

Carlo Rega

Abstract Landscape preservation and development of rural areas are two fundamental European Union's policy objectives towards sustainable development, as defined, respectively, by the European Landscape Convention and the Rural Development Policy, the second pillar of the Common Agricultural Policy (CAP). Agriculture has been shaping the landscape for millennia in Europe and with more than 47 % of EU territory devoted to it, the predominant form of landscape is indeed the rural one. Whilst the deep interlinks between rural development and landscape policies/planning are evident, these two domains have developed, both as research fields and policy sectors, largely independently. Traditionally, rural development policies have featured a sectoral approach, conceiving agriculture as an economic activity, without properly considering its territorial dimension; similarly, landscape and territorial planning have not paid adequate attention to the specific needs of agriculture and farmers in designing preservation measures. The challenges Europe faces in the 21st century towards the objective of sustainable development urge for a deeper integration of these two domains. This is particularly true after the adoption, in December 2013, of the CAP reform package comprising the new regulations on the Rural Development Policy for the programming period 2014–2020. This introductive chapter provides an overview of the key concepts and issues addressed in the text: the role of agriculture in shaping the landscape; the ecosystem service conceptual framework; the concept of landscape itself and rurality; the meaning of rural development and multifunctionality in agriculture and rural areas. Subsequently, a synopsis of the following chapters is presented.

Keywords Landscape planning · Rural development · Agriculture · Ecosystem services

---

C. Rega (✉)
Interuniversity Department of Regional and Urban Studies and Planning (DIST), Politecnico di Torino, Viale Mattioli 39, 10125 Torino, Italy
e-mail: carlo.rega@polito.it; carlo.rega@jrc.ec.europa.eu

C. Rega (ed.), *Landscape Planning and Rural Development*, SpringerBriefs in Geography, DOI: 10.1007/978-3-319-05759-0_1,

1.1 Agriculture, Landscape, Rural Development: An Overview of Key Definitions and Concepts

No other activity testifies the deepness and complexity of the interrelations between nature and humankind like agriculture. Its introduction in Europe dates back to 8,000 years ago and allowed the transition from nomad societies, based on hunting and gathering, to permanently settled ones—the so-called Neolithic revolution. It was agriculture that posed the basis for human population growth and the development of cities, as well as trading, made possible by the food surplus generated by permanently cropped areas. Not surprisingly, the first important civilizations like ancient Egyptians, Assyrians and Babylonians, all developed in fertile areas like the Nile floodplains or the Fertile Crescent. In more recent times, the industrial revolution itself, started in England at the end of the 18th century, was driven by a previous agricultural revolution that had significantly increased yields and food production through the introduction of innovative management systems, like the four field crop rotation.

Agricultural activity is made possible by the constant modifications that men produce on natural habitats with the aim of improving working conditions and yields. In ecological terms these new areas—agroecosystems—can be defined as (natural) ecosystems that have been deliberately modified by humans for purpose of the production of specific species of value to them (see e.g. Swift et al. 2004). Such transformations always imply a simplification of the original ecosystems and a decrease in biodiversity at all levels—genetic, species and habitat. This is obtained by altering the original fluxes of energy and matters through ecosystems by adding external energy inputs (in the form of labour, heath, water, chemicals) in order to maximize the conversion of such energy into biomass that has a direct value for humans—food, fibre, feed—and minimize the conversion to non useful biomass (e.g. weeds). By doing so, these modifications also affect virtually all features of natural ecosystems: the very morphology of the terrain (e.g. levelling, terraces); its composition and texture (through ploughing, fertilization); the hydrography (construction of ditches, canals); the hydrological cycle (water consumption for irrigation, increased storage of water in biomass); the microclimate (through e.g. alterations in evapotranspiration, air humidity) and so on.

From an anthropocentric perspective, most of the natural processes at the base of the functioning of (agro) ecosystems can be conceptualised in terms of *ecosystems services*, defined as "the direct and indirect contributions of ecosystems to human well-being" by the Economics of Ecosystems and Biodiversity report (TEEB 2010). Following the well-established taxonomy firstly elaborated by the Millennium Ecosystems Assessment (MEA 2005) and subsequently refined by TEEB, four categories of services are identifiable: provisioning (e.g. food, fresh water, wood and fibre, fuel); regulating (e.g. climate regulation, flood and disease regulation and water purification); habitat (e.g. as nurseries and gene-pool "protectors"); and cultural and amenity services (aesthetic, spiritual, educational and recreational).

Agricultural areas are at the same time *consumers* and *providers* of ecosystem services. On the one hand, they rely on services provided by natural ecosystems such as pest control, pollination, regulation of water quantity and quality, and soil fertility (Power 2010; Swift et al. 2004; Zhang et al. 2007). On the other hand, and depending on the management system adopted, agricultural areas can provide some services and functions (beyond, of course, provisioning ones), such as carbon sequestration, habitat provision, pollination, flood mitigation, regulation of soil fertility, nutrient cycling (Maes et al. 2011; Power 2010; Rega and Spaziante 2013). Rural areas have also a key role in providing cultural and amenity services, e.g. by offering the opportunity for outdoor recreation and (agri) tourism.

The conceptual framework provided by TEEB is useful to the purposes of this book because it clearly distinguishes between ecosystem structures/processes, functions and services. Ecosystem structures and processes interact in physical (e.g. infiltration of water), chemical (e.g. oxidation) or biological (e.g. photosynthesis) ways; a subset of these interactions are ecosystems functions, such as production of biomass, water regulations, habitat provision, and information. These functions in turn may be used (directly or indirectly, intentionally or unintentionally) by men in the form e.g. of food production, flood prevention, nursery or recreation. By making this distinction, the TEEB framework focuses on the humans' point of view in their relation with the environment, which is crucial also to conceptualize another key term in the context of this book, namely *landscape*.

All the efforts that humans have been making over the last millennia to convert natural ecosystem into agro-ecosystem have been one of the main driver in the shaping of what we commonly refer to as "landscape". This term may be intended with slight different meanings by different fields of study: here we refer to the definition given by the European Landscape Convention (ELC): "an area, as perceived by people, whose character is the result of the action and interaction of natural and/or human factors" (CoE 2000, art. 1). This definition highlights two elements: the first one refers to the human perception and, as it was previously anticipated, it is crucial because it allows to distinguish the concept of landscape from other ones such as "environment", "territory" or "habitat" (see also Cassatella and Seardo, *infra*, Chap. 3), often used synonymously, which may create ambiguity. A portion of a territory is a part of the environment and may comprise one or more habitats, but what makes it a landscape is the fact that is perceived in a certain way by people, just like an ecosystem function becomes a service only if it is used by men for their well-being. The second one is that, in turn, *each portion of the territory is a landscape* as far as it is perceived by people, regardless of its appreciation in terms of attractiveness, openness, rarity, socio-cultural importance and so on. This is relevant because it clarifies that landscape policies shall not be limited to the preservation of outstanding landscapes but also to the improvement of more "ordinary" ones.

The ELC definition strongly resonates with the one that Emilio Sereni, 40 years before, gave of *rural landscape*: "the shape that people, by their agricultural activities, systematically and consciously give to the natural landscape" (Sereni

1961). The author, in its seminal work on the history of Italian agricultural landscape, pointed out that this process was not only the results of bio-physical conditions and evolution of agricultural techniques, but also of the constantly evolving economic and property relations between different social agents, as workers and landowners. Social struggles played a crucial role in transforming the landscape: to make but one example, think to the repossession processes of latifundium by small farmers after World War II in Italy (and in other countries as well), with the consequent replacement of fallow lands and increase of fields fragmentation. This is important in the context of this book to dispel possible misunderstandings or somewhat naive views of rural areas as idyllic places far from the tensions and contradictions of metropolitan areas. Far from being a static concept, rural landscape is even more than a dynamic one: it can be indeed considered as a *dialectic* space, the resultant of different and often contrasting drivers.

The world land area dedicated to agriculture is 38.5 % of the total emerged area (FAO 2014), this figure rising up to 43 % in the EU27 (EUROSTAT 2014), where woodlands account for approximately 39 %, artificial areas for 4.3 % and waters for 3.4 % of the total (EUROSTAT 2014). Hence, rural landscape is one of the predominant forms of landscape worldwide and particularly in Europe, and one that virtually everyone has experience of. Yet, the term *rural* has no conventional and world-accepted definition. Ashley and Maxwell (2001) define rural space as a "space where human settlements and infrastructure occupy only small patches of the landscape, most of which is dominated by fields, pastures, woods, waters, mountains and deserts". In the European context, the European Conference of Ministers responsible for Spatial/Regional Planning (CEMAT) defines Rural areas/countryside as "sparsely settled areas without significant large city or town. The countryside refers to certain forms of landscapes and land uses where agriculture and natural areas play an important part" (CEMAT 2007, p. 23). Ekong (2010, as cited by Adisa 2012) identifies the main features distinguishing rural settings from urban ones, namely smaller size of communities; lower population density, greater proximity to nature, higher occupation rates in agriculture and related activities, as well as more sociological differences like lower social mobility, income differentiation and standard of living. Whilst most of these characteristics can indeed be found in rural areas around the world—particularly in developing countries—they do not apply to all contexts and periods.

The Organisation for Economic Co-operation and Development (OECD 2001) defined a method to identify urban-rural territories based on population density, establishing a threshold of 150 inhabitants per square kilometres as the upper limit for rural areas. This criterion was first adopted by the European Union and applied to each level 2 local administrative units (LAU2—the smallest administrative unit defined by the European Union, corresponding to municipalities in most countries). Larger administrative units (NUTS3) were classified as predominantly urban, intermediate and predominantly rural if the share of population living in rural LAU2 was below, respectively 15 and 50 %. Subsequently the methodology was refined to correct distortions by defining a population density threshold of 300 inhabitants per km^2, applied to grid cells of 1 km^2, plus a minimum size threshold of 5,000 inhabitants applied to grouped grid cells above the density threshold. The threshold separating predominantly urban and intermediate areas was also increased from 15 to 20 %. According to this new classification method, predominantly urban, intermediate and predominantly rural areas account respectively for 9.1, 34.2 and 56.7 % of the EU 27 territory. In terms of population, these figures change respectively to 41.0, 35.5 and 23.6 %. Whilst all definitions based on population density have the advantage of objectivity, they clearly cannot convey the complexity and different traits of "rurality", which, in turn, vary considerably between cultures and countries and even between regions within the same country. Even if this is not optimal from a research perspective, we probably have to acknowledge that the term rural inevitably bear a certain degree of ambiguity and is subject to different interpretations.

Given this, it is even more difficult to define the concept of *rural development*. Until the seventies, this term was understood as synonymous of agricultural development. In the eighties, the World Bank defined it as a set of strategies and policies to improve the economic and social life of rural poors (Adisa 2012). Just like the concept and consideration of rurality greatly varies depending on the contexts and periods, so rural development can encompass a variety of meanings: in developing countries it is mainly related to poverty alleviation, nutrition, improving of health conditions, illiteracy eradication, access to natural resources and community services (Adisa 2012). In developed countries, the concept is more linked to aspects such as quality of life and environment preservation: the US Department of Agriculture for instance defines rural development as "improvement in the overall rural community conditions, including economic and other quality of life considerations such as environment, health, infrastructure, and housing" (USDA 2006).

In the European Union, the main objectives of the Rural Development Policy as established by Regulation 1305/2013 are (i) fostering the competitiveness of agriculture; (ii) ensuring the sustainable management of natural resources, and climate action; and (iii) achieving a balanced territorial development of rural economies and communities including the creation and maintenance of employment. This policy is framed into the broader Common Agricultural Policy (CAP), one of the EU's most long standing and important ones in terms of budget and scope. The CAP's objectives have undergone profound changes since its inception: originally envisaged in the aftermaths of World War II as a mean to support agricultural production and prices, ensure food self-sufficiency in Europe, and create a common market, it has evolved over the time to encompass broader objectives, such as diversification of rural activities and environmental protection. A significant turn was prompted in 1985 by the issue of the "green paper" perspective of the Common Agricultural Policy (CEC 1985). Here for the first time the need for agriculture to take environmental issues into account was explicitly stated: "the role of agriculture in a modern industrialized economy is increasingly perceived to include not only the strategic, economic and social functions [...], but also the conservation of the rural environment" (Adisa 2012, p. 49). It was also acknowledged that agriculture, like other economic sectors, had potentially harmful activities that shall be subject to control and limitations.

This publication fuelled a strong debate that lead to the so-called Mac Sharry reform in 1992, which established a set of measures specifically aimed at mitigating the negative environmental impacts of agriculture and supporting environmental friendly practices through the introduction of agri-environmental schemes (see Rega, *infra,* Chap. 2). But only for the 2000–2006 programming period an autonomous and organic Rural Development Policy was defined as integral part of the CAP (known as its "second pillar", the first one being support to production). For the 2007–2013 programming period, the EU Rural Development Policy was articulated in three main axes: (i) improving the competitiveness of the agriculture and forestry sectors; (ii) improving the environment and countryside; (iii) rural quality of life and diversification of the rural economy.

Following the above-mentioned CAP reform, for the programming period 2014–2020 the three axes will be replaced by six *priorities*:

- fostering knowledge transfer and innovation in agriculture, forestry, and rural areas;
- enhancing farm viability and competitiveness;
- promoting food chain organisations;
- restoring, preserving and enhancing ecosystems related to agriculture and forestry;
- promoting resource efficiency and supporting the shift towards a low carbon and climate resilient economy;
- promoting social inclusion, poverty reduction and economic development in rural areas.

As part of the CAP, the EU Rural Development Policy is called to contribute to the achievement of the Europe 2020 objectives of smart, sustainable and inclusive growth. Rural development is conceived as a multidimensional and integrated concept, premised upon *multifunctionality* and *diversification*, with a strong emphasis on the environmental dimension and the socio-economic development of rural communities.

As other concepts we have presented in this introductory chapter, *multifunctionality* is a controversial one (Daniel and Perraud 2009; Cassatella and Seardo, *infra*, Chap. 3). Perhaps the most cited definition is the one provided by the OECD (2001, p. 7): "The key elements of multifunctionality are: (i) the existence of multiple commodity and non-commodity outputs that are jointly produced by agriculture; and (ii) the fact that some of the non-commodity outputs exhibit the characteristics of externalities or public goods, with the result that markets for these goods do not exist or function poorly". A recurring example provided by the OECD report to illustrate non-commodity public good is precisely landscape. The report acknowledges how agriculture affects and shapes landscape through a combination of decisions regarding land use, farming practices, commodity compositions and many others. The association of landscape with particular production systems (e.g. terraced fields, alpine pastures, vineyards), is also cited as a paradigmatic example of joint products caused by non-allocable inputs within the farms (OECD 2001). As rural development measures (and particularly agri-environmental schemes) directly affect all these aspects (see Rega, *infra*, Chap. 2), it is clear that a strong link exists between rural development and landscape preservation/improvement, even when the latter is not an explicit objective of such schemes.

This makes the case for the deeper integration between rural development policies/tools and the policies and tools explicitly aimed at landscape preservation, which is the overarching argument of this publication. At this point, a sharper definition of terms such as *landscape planning* and *landscape policies* is thus needed. Again, no single, world-wide accepted definition probably exists, but in the context of the EU we can refer to the one provided by the CEMAT (2007): "Landscape planning is an activity involving both public and private professionals, aiming at the creation, conservation, enhancement and restoration of landscapes at various scales, from greenways and public parks to large areas, such as forests, large wilderness areas and reclamation of degraded landscapes such as mines or landfills. Landscape planning encompasses a variety of skills, such as landscape architecture and design, nature conservation, knowledge of plants, ecosystems, soil science, hydrology, cultural landscapes, etc." As for landscape policies, the CEMAT refers to the definition provided by the ELC (CoE 2000, art. 1): "[…] an expression by the competent public authorities of general principles, strategies and guidelines that permit the taking of specific measures aims at the protection, management and planning of landscapes".

Based on these conceptual premises, the next section illustrates the objectives and structure of the publication.

1.2 Aims and Structure of the Book

The overview presented in the previous section has touched upon the main concepts that readers will come across in the following chapters. The aim of this book is to address some of the main aspects outlined above through concrete case studies. It does not pretend to be a thorough compendium of all the issues related to landscape and rural development but it seeks to present significant experiences and practices as well as operational tools. Authors are scholars and practitioners with different backgrounds and the book in turn intends to offer useful insights either to scholars, practitioners and policy makers, as well as to students. The case studies presented are all from a single country—Italy—but cover a variety of landscapes and territorial contexts that can be of interested for virtually all European countries and beyond, including mountain sites, intensive agriculture zones, characteristic agricultural landscapes, and peri-urban areas.

Following on the same argument presented in the previous section, in Chap. 2 Carlo Rega engages with the issue of integration between landscape/spatial planning and Rural Development Programmes (RDP), with a focus on

Agri-Environmental Schemes (AES). First, the chapter provides an overview of current (2007–2013) and forthcoming (2014–2020) RDP objectives and contents. Subsequently, shortcomings in AES implementation are identified and discussed, with a focus on spatial targeting and coordination between beneficiaries, which makes a strong case for the argued integration with spatial/landscape plans. Theoretically, a distinction between two different rationales—termed *sectoral* and *territorial*—underlying landscape planning and RDP is presented and discussed. The author argues that the first one strongly permeates current RDP and AES design and implementation, conceiving agriculture mainly as an economic activity. The territorial dimension instead conceptualises agriculture as one of the main activity shaping the territory and the landscape, and stresses its interlinks with landscape and environmental preservation. The argument proceeds by showing that landscape plans generally feature a more detailed analytical apparatus and set rules and visions for landscape management with a holistic approach. On the other hand, their implementation is mainly indirect and they generally cannot fund projects. Conversely, RDP directly affect landscape through funding concrete actions, but their overall effect is limited by a sectoral focus, and by the lack of coordination between single agents undertaking these actions (farmers). The case study of the Piedmont RDP and the Turin Provincial Territorial Plan (PTP) is then presented: synergies and convergences between the policies and objectives of these two instruments are identified, and the territorial/landscape classification provided by the PTP is used to refine the identification of priority areas on which to target two major AES through spatial analysis in a GIS environment.

The exercise supports the argument of a shift towards what is defined a *territorial governance* approach to AES. The main features of such an approach are discussed, the argument being that beyond more spatially targeted AES, also joint projects based on the identification of local stakeholders and partnerships can be elaborated so to mobilise the local knowledge and capacities, with the overall aim of increasing their effectiveness. The implications of this paradigm shift are discussed to point out the need of increased horizontal and vertical coordination between public bodies and policy sectors, as well as deeper involvement of civil society organizations on a territorial base. The proposal echoes recent works from different research fields, and intends to contribute to the forthcoming implementation of the new European Rural Development Regulation, and as a way to pursue the implementation of the European landscape Convention and the EU objectives of sustainable development.

As already mentioned, multifunctionality is a key—yet controversial—concept for exploring the landscape/rural development interface. In Chap. 3, Claudia Cassatella and Bianca Seardo critically discuss it with reference to cultural ecosystem services and in particular scenic/aesthetic values. The authors call into question the assumption that environment—and landscape-oriented actions will always coincide or otherwise derive mutual benefits, and warn against addressing landscape issues only by means of strictly environmental indicators derived from landscape ecology or land uses. As a consequence, a deeper consideration of the aspects that distinguish landscape from environment, as scenic ones, is needed, both theoretically and as regards operational tools. The second part of the chapter illustrates a methodology for the assessment of scenic landscape, articulated in three main steps: (i) recognition and representation of scenic characteristics; (ii) visual analysis through Geographic Information Systems; and (iii) development of visual landscape indicators. The methodology is applied in three case studies in Northern Italy featuring different landscapes: a protected rural landscape (Albugnano, surroundings of Vezzolano Abbey), a landscape in transition (San Martino Alfieri), and a peri-urban landscape (the metropolitan area of Turin). Through these cases, it is shown how scenic assessment can fruitfully inform planning and decision making and identify possible trade-offs. Findings support the idea that integrating rural, landscape and spatial policies is necessary, and that such integration can be achieved through sharing knowledge apparatus, strategic objectives and assessment frames and indicators.

In Chap. 4 Chiara Bragagnolo, Chiara Rizzi and Stefania Staniscia present a multi-scale approach to support integrated landscape management in Rural Mountainside Areas of Alps, funded by the Autonomous Province of Trento, a mountain region in northern Italy. The presented methodology includes the definition and localisation of Rural Mountainside Areas, the selection of priority contexts and the definition of integrated measures to enhance identified values and mitigate risks. It combines top-down, GIS-based analyses with bottom-up approaches—expert opinion and field studies. Based on these analyses, the authors propose three main strategies for enhancing rural landscapes of Rural Mountainside Areas, namely: protection and conservation, tourism development and environmental sustainability. Then, three targets for implementing the proposed strategies are identified: rural buildings and assets, production systems, and accessibility. Finally, by crossing strategies with targets, a set of integrated operative measures is proposed. The presented case study shows, on the one hand, that the integration of multiple expertises is key in elaborating landscape preservation policies in rural areas, and, on the other hand, that the proposed measure for enhancing landscape management, including the definition of mechanisms and incentives, can be easily integrated into the Rural Development Program. However, the authors also point out that rural development policies are still characterised by a sectoral approach and that the advocate integration need to overcome the scarce coordination among different Offices of the Authority (*in primis*: Agriculture and Regional and Town Planning and Landscape Protection) as well as between different administrative levels (which resonates with similar considerations by Rega, *infra*, Chap. 2). To facilitate this cooperation, political willingness seems imperative.

As pointed out in the previous section, central to the conceptualization of landscape as defined by the ELC is the aspect of human perception. It therefore appears fundamental that any landscape strategy, policy or plan be based on studies and empirical evidence that explicitly take into account the (stated or implicit) preference of people. Chaps. 5 and 6 address this issue by presenting operational tools and techniques in different landscape contexts.

In Chap. 5, Marta Bottero presents an application of economic valuation techniques to assess the economic value of landscapes in a touristic Alpine valley in Piedmont, Northern Italy. The method is based on Conjoint Analysis, a set of statistical methodologies which aim to study individual choices using preferences expressed about various profiles. The tool is applied to survey the preferences of people in relation to different landscape components—land use, openness, repetitiveness, naturalness, and cost—and to assign them a monetary value. The method entails the submission of a questionnaire with questions regarding people's attitudes towards the particular landscape under analysis, the specific conjoint analysis questions, and socio-economic information about respondents. Findings indicate that the most influent variable on the respondents' preferences is naturalness (positive correlation), followed by the price and the presence skiing facilities (negative) and landscape uniqueness (positive). From this, willingness to pay and implicit prices of different landscape elements are calculated, to evaluate the degree to which the interviewees do trade-offs among the attributes. By showing that people do assign economic values to specific landscape features, and by calculating those values, the results of the exercise could be useful to inform and orient policy making and different level. As regards spatial/land-use planning, they can indicate to planners and local decision makers the most unwanted land use changes (as construction of skiing facilities); in the frame of rural development policy, they provide benchmark prices that can be used, for instance, to determine the amount of premium paid to farmers for the preservation of landscape features on their land. Despite some recognised limitations of the method, Conjoint Analysis has the advantage of using a *problem structuring* approach, based on "values focus thinking". This allows to identify desirable decision opportunities and alternatives opposite to conventional decision making that has traditionally focused on the evaluation of readily available alternatives, implicitly assuming that all decision problems are well structured. This appears particularly relevant in the case of complex decision making contexts, as is the case for landscape planning and rural development.

Rinaldo Grittani, Alessandro Bonifazi and Andrea Tassinari report in Chap. 6 about a participatory application of Landscape Character Assessment (LChA) to peri-urban country in the metropolitan area of Bari, southern Italy. Their motivation is rooted in an interest for putting two innovative approaches to landscape research and policy making to the test. Following the principles laid down in the ELC, they embark in a participatory evaluation of landscape quality and policies, choosing peri-urban space to epitomise everyday landscapes—about which the ELC calls for an adequate consideration, on a par with places of outstanding beauty. The authors provide a broad conceptual background, dwelling on three key aspects. Firstly, they elaborate on how "the concept of landscape proposed by the convention implies an exercise in democracy" (Council of Europe 2008: par. II.2. 3.A) by linking public participation in landscape policy making, the reconfiguration of expert-citizen relationships under research-action approaches, and the role of social practices in re-producing landscapes. Then, a reconsideration of landscape quality is attempted at, by focussing on photography in visual analyses to grasp the influence of perceptual aspects. The conceptual background ends with a discussion of peri-urbanity as a hybrid space that blends (neo) rural traits with contemporary urban dynamics—including sprawling. A case study is presented, based on the implementation of an adapted LChA protocol in a small town—embedded in a metropolitan agglomeration—where declined agricultural activities gave way to urban sprawl. The authors adopted a mixed-method design, where desk studies and field surveys are complemented by interactive social research methods (including semi-structured interviews, rural walks and focus groups) to effectively involve local stakeholders. Findings point to peri-urban countryside as being an elusive subject for LChA, and highlight the frictions that arise when public participation is integrated into expert-driven approaches. Under LChA, landscape characterization results in a map that closely resembles zoning instruments and is as such perceived by citizens, who accordingly tend to focus debates on development rights and the opposition between urbanization and conservation. Moreover, issues of scale aren't easily handled by local communities, who in turn seem to be the ideal trigger for landscape participatory governance. Finally, the main contribution that the Chapter is likely to add to the existing literature lies in strengthening the argument for imaginative strategies to cope with the in-betweenness of peri-urban landscapes, as well as in bringing neo-rural social practices to the fore as a means to reconcile rural development programmes and landscape policy.

The different contributions aim to offer evidence-based policy recommendations; they do not have the pretension of presenting necessarily "best" practices but rather aim to shows concrete options and examples, as well as limitations and difficulties, of what can be done at the interface of landscape planning and rural development. The broader horizon is the implementation of both the new EU Rural Development Policy and the European Landscape Convention. As highlighted at the outset of this section, these represent two important European policy objectives towards the achievement of the Europe 2020 strategy, particularly with regard to the major environmental challenges humanity has to face: climate change, biodiversity conservation, depletion of natural resources and the related socio economic ones—income distribution, inequalities, employment, and migration. This book aspires to make a modest but useful contribution in this sense.

References

Adisa RS (2012) Introductory chapter—rural development in the twenty-first century as a global necessity. In: Adisa RS (ed) Rural development—contemporary issues and practices. InTech, Rijeka, pp 3–14. doi: 10.5772/1399

Ashley C, Maxwell S (2001) Rethinking rural development. Dev Policy Rev 19(4):395–425

CEC (Commission of the European Communities) (1985) Perspectives for the common agricultural policy (Green paper). EC Official Publication, Luxembourg

Council of Europe (2000) European landscape convention. Council of Europe, Florence

Daniel F-J, Perraud D (2009) The multifunctionality of agriculture and contractual policies. A comparative analysis of France and The Netherlands. J Environ Manage 90(suppl. 2):S132–S138

Ekong EE (2010) Rural sociology. Dove Educational Publishers, Uyo, pp 380–382

European Conference of Ministrers Responsible for Spatial/Regional Planning (CEMAT) (2007) Spatial Development Glossary / Glossaire du développement territorial (bilingual). (Territory and Landscape Series No. 2). Strasbourg, Council of Europe

Eurostat (2014). Agricultural statistics. http://epp.eurostat.ec.europa.eu/portal/page/portal/agriculture/introduction. Accessed 3 Feb 2014

FAO (Food and Agriculture Organisation of the United Nations) (2014) Agri-environmental statistics. http://www.fao.org/economic/ess/ess-agri/en/. Accessed 2 Feb 2014

MA (Millennium ecosystem assessment) (2005) Ecosystems and human well-being: current state and trends assessment. Island Press, Washington

Maes J, Paracchini ML, Zulian G (2011) A European assessment of the provision of ecosystem services. Towards and atlas of ecosystem services. EUR 24654 EN—Joint Research Centre—Institute for Environment and Sustainability. Publications Office of the European Union, Luxembourg. doi:10.2788/63557

OECD (Organisation for Economic Co-Operation and Development) (2001) Multifunctionality: towards an analytical framework. Agr food, Paris

Power AG (2010) Ecosystem services and agriculture: tradeoffs and synergies. Philos Trans R Soc Lond B 365:2959–2971

Rega C, Spaziante A (2013) Linking ecosystem services to agri-environmental schemes through SEA: A case study from Northern Italy. Environ Impact Assess Rev 40:47–53

Sereni E (1961) Storia del paesaggio agrario italiano, Editori Laterza, Roma-Bari

Swift MJ, Izac AMN, van Noordwijk M (2004) Biodiversity and ecosystem services in agricultural landscapes—are we asking the right questions? Agric Ecosyst Environ 104:113–134

TEEB (2010) The economycs of ecosystems and biodiversity: Mainstreaming the economics of nature: a synthesis of the approach, conclusions and recommendations of TEEB

USDA (2006) United States Department of Agriculture—2007 Farm Bill Theme Papers Rural Development Executive Summary. Available online at http://www.usda.gov/documents/Farmbill07ruraldevelopmentsum.pdf

Zhang W, Ricketts TH, Kremen C, Carney K, Swinton SM (2007) Ecosystem services and disservices to agriculture. Ecol Econ 64:253–260

Chapter 2
Pursuing Integration Between Rural Development Policies and Landscape Planning: Towards a Territorial Governance Approach

Carlo Rega

Abstract Landscape planning and Rural Development Programmes (RDP) share common objectives of preservation and improvement of the rural landscape, so a deeper integration between these two domains would deliver significant benefits towards sustainable development. However, until now they have developed largely independently as both research fields and policy sectors. This chapter addresses the main theoretical issues concerning the advocated potential integration by first identifying and discussing two different rationales underlying landscape and rural development policies, namely a *territorial* and a *sectoral* one. Subsequently, a case study regarding the Territorial Plan of the Province of Turin, Italy, is presented to illustrate how landscape/spatial planning and RDP's policies and objectives can converge and the different regulations and capacities of these instruments used to deliver mutual benefits. In particular, it is shown how the design and implementation of Agri-environmental schemes within RDP could be made more effective and spatially targeted by taking into account the spatial analysis and landscape areas designation elaborated by the Territorial Plan. It is argued that to foster synergies, a shift towards a *territorial governance* approach in RDP design and implementation is needed, which entails a deeper horizontal and vertical coordination between government levels and sectors, as well as the involvement of stakeholder from the civil society in the design and realization of territorialized projects. The discussion is framed in the ongoing debate on the implementation of the Common Agricultural Policy in the period 2014–2020.

Keywords Landscape planning · Rural Development Programmes · Agri-environmental schemes · Territorial governance

---

C. Rega (✉)
Interuniversity Department of Regional and Urban Studies and Planning (DIST),
Politecnico di Torino, Viale Mattioli 39, 10125 Torino, Italy
e-mail: carlo.rega@polito.it; carlo.rega@jrc.ec.europa.eu

C. Rega (ed.), *Landscape Planning and Rural Development*,
SpringerBriefs in Geography, DOI: 10.1007/978-3-319-05759-0_2,

2.1 Introduction

Landscape preservation and rural development are two pivotal policy objectives in the European context as defined, respectively, by the European Landscape Convention—hereafter ELC—(CoE 2000) and the recently approved EU Regulation on support for rural development by the European Agricultural Fund for Rural Development (EAFRD) (EU 2013) n. 1305/2013, replacing previous regulation n. 1698/2005. The ELC acknowledges the importance of agricultural activity in shaping the landscape (ELC, preamble) and provides that each Member State shall undertake "to integrate landscape into its regional and town planning policies and in its cultural, environmental, *agricultural*, social and economic policies [...]" (art. 5 d, emphasis added), thus explicitly highlighting the interrelation between landscape and agriculture.

Landscape policies, however, fall under the responsibility of Member States and there is no specific legislation on landscape management in the legal frame of the European Union (EU). Nonetheless, EU policies do affect landscape, and this is particularly the case for the Common Agricultural Policy (CAP), since agricultural land covers 47 % of the EU-27 surface, this figure increasing to 78 % if forests are included. Therefore, measures implemented in the frame of the CAP and its second pillar—rural development—directly affect a significant share of the EU territory (Paracchini and Capitani 2011; Lefebvre et al. 2012). In this context, landscape has become a central element in the CAP debate (Vergamini et al. 2013).

The main argument put forward in this chapter is that, despite different rationalities and regulatory frames underlying landscape planning and rural development policies, these two domains can and shall be more integrated, and that such integration would be mutually beneficial for the achievement of their respective objectives and towards sustainable development as outlined in the Europe 2020 Strategy. The aim of this chapter is to present and discuss the key issues and challenges that such integration poses in the context of the EU, particularly following the recent CAP reform, and to present a case study providing a practical example of how this could be achieved. The chapter is structured as follows: in Sect. 2.2, the main characteristics of Rural Development Programmes (RDP) are presented; current shortcomings concerning their implementation in the period 2007–2013 are also pointed out, with particular reference to agri-environmental schemes and their potential impact on landscape. The main innovations of the CAP reform are presented and discussed, with a focus on the Rural Development Regulation. In Sect. 2.3, the differences in the rationales underlying landscape planning and RDP are identified and discussed. In Sect.2.4, a case study concerning the Territorial Plan of the Province of Turin (Italy) and the regional RDP is presented to show possible options to concretely implement the advocated integration. In Sect. 2.5, the main points emerged from the theoretical sections and the case study are summarized and discussed. Section 2.6 concludes.

2.2 Rural Development Policy in Europe: An overview

2.2.1 Rural Development Programmes and Agri-Environmental Schemes: Key Features and Issues at Stake

The Common Agricultural Policy (CAP) is one of Europe's most long standing and relevant ones, in terms of both spatial coverage and expenditure—it will account for about 38 % of the whole EU budget in the period 2014–2020 (EU 2013). Measures implemented under the CAP deploy their effects on a large share of the European territory and therefore can have a significant influence in the preservation or depletion of landscape (Paracchini and Capitani 2011). It is widely acknowledged that the CAP itself has been a major driver of environmental and landscape degradation, due to the incentive to intensive farming management aimed at maximizing production (Garrod 2009). The CAP comprises two main policies: (i) economic support to production and (ii) Rural Development Policy. Known as the "second pillar" of the CAP, the latter aims to improve the efficiency of the agricultural and forestry sectors while preserving and enhancing the environment and the landscape. Already in 2010, the EU indicated the key challenges for European agriculture in the document "The CAP Towards 2020: Meeting the Food, Natural Resources and Territorial Challenges of the Future", namely: guaranteeing food security, preserving natural assets and the rural landscape, combating climate change, and pursuing territorial cohesion and balance (EC 2010). By responding to these challenges, the CAP shall also contribute to the implementation of the EU 2020 Strategy in terms of smart, sustainable and inclusive growth. The Rural Development Policy is implemented by Member States through the elaboration of Rural Development Programmes (RDP), which allocate funding to farmers and other land managers for the realization of the objectives set by the rural development regulation. In some countries, as Italy, Spain or Germany, RDP are elaborated at the regional level.

In the programming period 2007–2013, RDP were developed along three main axes: (1) Improving the competitiveness of the agriculture and forestry sectors; (2) Improving the environment and countryside; (3) Rural quality of life and diversification of the rural economy. The funding supported farmers and other stakeholders for additional costs they incur for the improvement of their work or as compensation for foregone income due to the implementation of more environmentally friendly farming practices.

Each axis comprised a certain number of measures, some of which in turn further articulated in sub-measures or actions. Each measure/action had a set of specific objectives deriving from the three general ones mentioned above. RDP do not directly determine the single projects to be realised, but, through the provision of funding, set the framework for the outcomes to be achieved, and the activities to be supported towards the attainment of those outcomes. In fact, funding is granted to beneficiaries generally through a system of calls for applications, periodically issued by the Managing Authority. In most cases, applicants are farmers but, depending on the measure, they could also be landowners, public sector organisations or rural communities. If the application is approved, the applicant receives funding to implement projects/measures that contribute towards the programme's stated objectives.

Since a certain budget is allocated to each measure/action, a system of scores is often used by the Managing Authorities to rank and select the applications that will be granted the subsidy, if the demand exceeds the available funds. Depending on the measure/action, priority scores may regard the nature of the applicant (e.g. young farmers) and/or the location of farms; in particular, for environmental measures, priority is given to certain areas identified based on their vulnerability or natural value, like Nitrate Vulnerable Zones or Natura 2000 sites.

Agri-environmental schemes (AES) are one of the main components of RDP. They are packages of actions voluntarily implemented by farmers in return for them adopting more environmental farming practices such as fertiliser and pesticide reduction, crop rotation, set aside areas, livestock extensification, and/or the creation, maintenance or enhancement of semi-natural elements in agro-ecosystems, as grass covers, hedgerows, ponds or tree lines. Contracts generally have a duration of at least five years and premiums cover only those additional commitments not already mandatory by law (referred to as "good agricultural and environmental conditions"—GAEC). Since these actions determine a loss of yield and/or increased costs for farmers (including transaction costs), premiums are determined by managing Authorities to compensate for such loss. The overall objectives of AES are to support the sustainable development of rural areas and respond to society's increasing demands for public environmental goods and services—including landscape—based on the assumption that public goods are not exchanged through market mechanisms, so farmers are not encouraged to supply them (EU 2013, considerandum n. 22).

For the programming period 2014–2020, the axes system will be replaced by six *priorities*, further articulated in *focus areas*, comprising voluntary measures, which, in the intention of the reform, would allow Member States to design the programs and their financing based on an analysis of their specific needs. These priorities are defined in Article 5 of the new Regulation 1305/2013:

(1) Fostering knowledge transfer and innovation in agriculture, forestry, and rural areas with a focus on the following areas:
    - fostering innovation, cooperation, and the development of the knowledge base in rural areas;
    - strengthening the links between agriculture, food production and forestry and research;
    - fostering lifelong learning and vocational training.

(2) Enhancing farm viability and competitiveness and promoting innovative farm technologies and the sustainable management of forests, with a focus on:

- improving the economic performance of farms and facilitating farm restructuring to increase market participation and agricultural diversification;
- facilitating the entry of adequately skilled farmers into the agricultural sector and, in particular, generational renewal.

(3) Promoting food chain organisation, with a focus on:

- improving competitiveness of primary producers by better integrating them into the agri-food chain through quality schemes, adding value to agricultural products, promotion in local markets and short supply circuits, producer groups and organisations and inter-branch organisations;
- supporting farm risk prevention and management.

(4) Restoring, preserving and enhancing ecosystems related to agriculture and forestry, with a focus on:

- restoring, preserving and enhancing biodiversity, including in Natura 2000 areas, and in areas facing natural or other specific constraints, and high nature value farming, as well as the state of European landscapes;
- improving water management, including fertiliser and pesticide management;
- preventing soil erosion and improving soil management.

(5) Promoting resource efficiency and supporting the shift towards a low carbon and climate resilient economy in agriculture, food and forestry sectors, with a focus on:

- increasing efficiency in water and energy use;
- facilitating the supply and use of renewable sources of energy, of by-products, wastes and residues and of other non food raw material;
- reducing green house gas and ammonia emissions from agriculture;
- fostering carbon conservation and sequestration in agriculture and forestry.

(6) promoting social inclusion, poverty reduction and economic development in rural areas, with a focus on:

- facilitating diversification, creation and development of small enterprises and job;
- fostering local development in rural areas;
- enhancing information and communication technologies in rural areas.

Restoring, preserving and enhancing European landscape is thus a stated objective of the Rural Development Policy, which makes a strong case for the pursuit of integration and synergies with the implementation of the ELC. Agri-environmental schemes (also called Agri-environment-climate payments in the new regulation) are one of the main tool to implement priority n. 4 and are detailed by art. 28. Despite the new architecture of the RDP, they maintain their main characteristics. The new regulation, however, also introduces some innovations that may foster integration with landscape planning, which are presented and discussed in Sect. 2.2.3.

2.2.2 Shortcomings in Current Implementation of Agri-Environmental Schemes in Europe

Over the last decade, AES have been the object of a large number of studies, given their relevance and their potential in steering agriculture towards more environmentally sustainable practices (Whittingham 2011). Overall, the empirical evidence collected suggests that they are still far from achieving their stated objectives, and that there is scope for improving their environmental performance and cost-effectiveness (Kleijn and Sutherland 2003; Mckenzie et al. 2013; Rega and Spaziante 2013; Whittingham 2011). Reasons for this are likely to be multifold, including problems with option design, monitoring difficulties, and the scale at which they are currently implemented (Mckenzie et al. 2013). The effectiveness of these schemes in fact depends on several factors, including the amount of economic compensations (Ruto and Garrod 2009), transaction costs (Mettepenningen et al. 2009), contracts' length (Lennox and Armsworth 2011) and spatial targeting (Uthes et al. 2010).

In 2010–11, the European Court of Auditors (ECA) carried out an audit in eight selected Member States and regions to examine the current implementation of agri-environmental schemes. The report ensued from this audit pointed out a series of shortcomings (ECA 2011):

- RDP objectives were not set out in a specific and measurable manner. The lack of quantitative thresholds makes it difficult to verify their achievement;
- Environmental pressures are often described in a general manner without providing a clear link with the AES;
- The common monitoring and evaluation framework is not geared to assess and measure the net environmental benefit determined by agri-environmental measures;
- Weaknesses were found in methods used to determine the amount of premiums paid for AES, which are also not sufficiently differentiated between regions with different conditions;
- For many area-based actions, a minimum uptake threshold has to be reached to produce tangible effect but despite this, Member States do not assess the minimum participation level needed to reach it;
- Implementation of AES is often not spatially targeted.

The last point deserve a closer look, as it makes a strong case for integration with landscape and spatial planning, as it will be argued in the following. The main idea of spatial targeting is that the positive net effect of AES is not an intrinsic element of measures themselves, but depends on the characteristics of the area where they are realized. To make an example, a decrease in nitrogen inputs from fertilization will produce a greater beneficial effect in a Nitrate Vulnerable Zone, other things being equal. Similarly, measures aimed at supporting biodiversity, as creation/upkeep of semi-natural vegetation, will maximize their effect if applied within, or nearby, high value ecological areas, like Natura 2000 sites. From an environmental point of view, a more efficient implementation would therefore imply to concentrate specific measures on identified areas, where their effect would be greater.

The scientific literature identifies poor spatial targeting as one of the major reasons for the lack of effectiveness of AES and advocates for its improvement (Piorr et al. 2009; Spaziante et al. 2013; Uthes et al. 2010; Vergamini et al. 2013). It shall be also noted that increased spatial targeting often imply higher private and public transaction and uneven distributional effects among farmers (Wunder et al. 2008). Furthermore, empirical findings also show that when multiple goals are assessed, not necessarily an increase in the degree of spatial targeting of a single measures improve the effectiveness of the program as a whole (Uthes et al. 2010). Overall, however, improved spatial targeting of AES is seen as a way to increase the environmental performance of RDP, as explicitly recognized by the cited report of the ECA (2011).

2.2.3 Main Element of the CAP and Rural Development Regulation Reforms

On December 16, 2013, the European Council adopted the CAP reform package. The reform sets out the new rules for the CAP in 2014–2020, its main stated objectives being to make the CAP greener and better targeted, more equitable and with a more effective rural development policy. The total CAP budget for the 2014–2020 period amounts to EUR 408.31 billion (38 % of the overall EU budget), of which 76.6 % will be devoted to the first pillar (direct income support and market-related expenditure) and the remaining 23.4 % will be allocated to the rural development policy. The CAP reform package comprises four main legal texts:

- the regulation establishing a common organisation of the markets in agricultural products;
- the regulation establishing rules for direct payments to farmers;
- the regulation on support for rural development;
- the regulation on the financing, management and monitoring of the CAP (known as horizontal regulation).

As regards the first pillar of the CAP, the main innovation in the context of this book is the so-called "greening" payment, according to which 30 % of direct income support for farmers will be granted only if they observe certain environmental farming practices. These include growing at least three different crops on arable land, maintaining a minimum area of permanent grassland, and preserving areas and landscape features with particular value, defined as "ecological focus area". These may comprise fallow land, terraces, buffer strips, agro-forestry, landscape features as hedgerows, tree lines, ponds and so on, and shall cover at least 5 % of the farms' agricultural area. Though subject to some criticism by advocates of a more environmentally oriented CAP reform (the requirement of ecological focus areas will only apply to farms with more than 15 ha of arable land) it is nonetheless evident how this innovation will affect European agrarian landscape.

As regards the rural development regulation, the main aspects of the reform concern:

- the enhancement of a common strategic framework, through the establishment of common rules on programming for all EU Funds and a deeper integration of the EAFRD with other European funds, with the objective of realizing innovative projects and coherent strategies in a given area, including urban-rural links;
- the replacement of the axis system with the above mentioned six priorities, which is intended as a way to simplify RDP's implementation and realization of projects by beneficiaries;
- The creation of a single measure to cover most types of physical interventions in farms, which is seen as a way to foster the realization of integrated projects;
- Simplification of the supporting scheme for investments in forestry, with a single integrated measure covering all investments;
- Introduction of a new delimitation of areas with natural constraints (previously defined as less favoured areas), to be implemented by Member States until 2018.

Measures regarding Agri-Environment (Article 28), Organic Farming (Article 29) and Natura and Water Framework Directive (Article 30), maintain the main characteristics of the previous period, with some innovations that in the intention of the reform should both increase their effectiveness and provide for higher flexibility—by allowing for shorter commitments periods and periods of conversion—in order to favour a wider uptake of the measures.

The main features of the now called Agri-environment-climate payment are the following:

- Their inclusion in national RDP is compulsory;
- Farmers and other land managers can participate on a voluntary base, individually or as groups;
- Premiums cover only those commitments that go beyond already mandatory standards;
- Commitments have a duration of at least 5–7 years, but where necessary Member States can determine longer or shorter periods;

- Premium are determined as a compensation for additional costs incurred, including transaction costs up to 20 % of the total amount paid (30 % if beneficiaries are grouped).

As regards selection criteria and spatial targeting, the regulation provides that Managing Authorities shall define selection criteria aiming to ensure equal treatment of applicants, better use of financial resources and targeting of measures in accordance with the Union priorities for rural development. However, this is not mandatory for agri-environmental-climate payment, organic farming and Natura 2000 and Water Framework Directive payment. In this latter case, eligible areas are agricultural and forest areas designated pursuant to Directives 92/43/EEC and 2009/147/EC and those concerned by the Water Framework Directive. As usual, premiums can cover only additional costs and income foregone resulting from disadvantages in the concerned areas and only in relation to commitments that go beyond already established GAEC. As in the previous period, beneficiaries may be selected based on calls for proposals, applying economic and environmental efficiency criteria.

2.3 Landscape Planning and Rural Development Programs: Different rationales and Potential Convergences

2.3.1 Different Rationales Underlying Landscape Panning and RDP

AES can have a direct impact on rural landscape, in some cases preservation of landscape features being one of their stated objectives. Whilst this reinforces the argument of a deeper integration with the instruments that implement landscape policies, i.e. landscape plans, a closer examination of the different rationalities and scopes of these two objects is needed.

Firstly, it is necessary to distinguish the different rationalities underlying the two pillars of the CAP. It was shown that obligations established by the direct payment regulation, notably the greening, might have similar effects on the landscape as some AES. However, in this case what is established is a minimum set of requirements farmers have to meet in order to get access to direct payments: some of such requirements establish the so-called GAEC, which are mandatory by law. Farmers do not receive financial support in return for them adopting GAEC, but comply with GAEC is a prerequisite for receiving funding under pillar I. If farmers do not comply with GAEC and, in the future, with greening provisions, they incur financial sanctions. What is applied here is thus the "polluter pay" principle: society acknowledges that agricultural activity may be harmful to the environment and requires farmers to comply with determined rules: this is the so-called cross-compliance.

Within pillar II, instead, support is granted through voluntary contracts stipulated between the public authority and individuals (farmers) who commit themselves to carry out environmental beneficial actions that go beyond what is mandatory by law in change of financial support. The rationale here is thus of a payment for a provided service deemed useful by the society, as preservation of landscape features or decrease of chemical inputs into the watershed.

Jeanneaux et al. (2011) define a classification of environmental services provided by agriculture that can be useful also to characterise different approaches to AES. They distinguish between: (1) internalization environmental services; and (2) provisioning environmental services. AES can be conceptualised as monetary fluxes paid for an environmental service provided in the frame of a transaction (ibid). This transaction takes place between the Managing Authority, representing the interests of the society, and individual farmers. The main point here is that agriculture areas provide public goods and services, but are generally privately owned. Since it is not feasible nor desirable to change the state of the ownership (to "expropriate" agricultural land), these contracts establish a way to guarantee that the provision of the service/good to the community is maintained, in the frame of a private property regime.

Here lies one of the main issues concerning what we refer to as the "mode of governance" of AES (we'll return to the concept of governance with more detail in Sect. 2.5): they are conceived as a remuneration for the provision of a good/service, but the amount of this remuneration is not proportional to the quality/quantity of the service/good provided. In fact, as said, premiums are defined as the sum of the additional costs and foregone income incurred by farmers committing to the AES. This poses two kinds of problems: the first one is the difficulty of calculating mean costs and foregone incomes, with distortions deriving from the application of such mean figures to a wide variety of situations, which lead to over or under pay farmers in many cases (as highlighted by the ECA report). The second problem, however, is more fundamental, as it concerns the very nature of this transaction between society and farmers: the service provided to the community greatly varies according to factors that are not strictly related to the costs farmers incur to generate them.

One of the most important factor is, as already mentioned, the environmental characteristics of the area enrolled to the scheme. To go back to the already proposed example, consider the decrease of N input on agricultural land. Less fertilizers application causes a certain decrease of yield, so the subside to the farmer is calculated by multiplying this decrease by the market price of the crop (foregone income), plus administrative and transaction costs (e.g. keeping a registry of all N application). Beyond technical difficulties and possible bias, the point is that the rationale to determine the support paid is completely based on the conception of agriculture as a purely economic activity. If the price of crop A is twice that of crop B, the farmers cropping A will receive twice the amount of the farmer cropping B for the decrease of N, other things being equal. The price the community pays to the farmers is not proportional to the benefits that it gets from them, which cause distortions and non-optimal levels of provisions of such services.

To further elaborate on this issue, a distinction shall be introduced here between two different rationales underlying the modes of governance of RDP, that we term, respectively, *sectorial* and *territorial* (see also Saraceno 2002 for similar considerations). The first one conceives agricultural mainly as en economic activity providing market goods and it's focused on the farmer as an economic agent, whilst the second has a broader perspective and considers also its environmental and landscape dimension, i.e. its interrelations with the rest of the territory. AES thus present a mix of both rationalities: the latter enters through the above mentioned system of priority scores used to target measure to more sensitive areas, or (as in the case of Natura 2000/Water framework Directive related payments) through the establishment of eligibility rules based on territorial criteria. However, empirical evidences shows that as regards implementation, it is the sectorial rationale that prevails, as spatial targeting policy prove not sufficient to determine spatially targeted outputs (ECA 2011; Spaziante et al. 2013). The sectoral rationale strongly permeates also all measures aimed to increase the productivity and efficiency of agricultural production, i.e. measures of former Axis 1 and now action related to priorities 1–3 as set by the new regulation. Conversely, measures of former Axis 3 and 4 and priority 6 of the new regulation, feature a predominantly territorial rationale, as they aim to the development of rural local communities with their distinguished and place-based mix of resources, based on perceived strengths and opportunities, and through the direct involvement of local stakeholders.

Turning again to AES, the benefits deriving from their implementation on the ground will greatly vary depending on farm's location. From an ecological perspective, the cumulative effects of AES are not simply additive, but are strongly path-dependent. This is particularly the case for all measure aimed at habitat preservation and support to biodiversity: it is a key principle of ecology that the capacity of a habitat to maintain its ecological functions is directly proportional to its area and inverse proportional to its perimeter. Put it simply, this mean that it would be preferable to have, say, a continuous area of two hectares dedicated to support biodiversity (e.g. set aside) instead of two separated areas of one hectare each. Path-dependency is also relevant for all AES regarding water pollution prevention, as the effects on a single portion of the land depend on what happens upstream in the water basin. Threshold effects is also an important factor: in many cases, no perceptible change in the environmental state occurs unless a specified farming practice is applied with a minimal intensity and on a minimal area in the zone of interest (Dupraz et al. 2009). Given the voluntary nature of AES, this again implies that concentration of actions would be preferable. The above examples serve to demonstrate that, for society as a whole, it is the territorial rationale the most important one in determining the importance of supporting farmers with AES.

The argument may be extended to landscape preservation as well: given the fact that preserving and enhancing the rural landscape through the maintenance of landscape features (hedgerows, ponds, trees, wet areas etc.) is a stated objectives of rural development, is may be reasonable that such actions are concentrated in areas deemed of particularly relevance from a landscape perspective, for instance, recognised and classified as such by landscape plans. As already clarified, such areas do not necessarily have to be "outstanding" landscapes, but may be as well more ordinary landscape, which might be recognised as strategic within a landscape policy, e.g. remnants of agricultural lands at the fringe of urban areas as a way to contrast urban sprawling and soil sealing. Again, this calls for a territorial approach to AES design and implementation.

2.3.2 How Landscape Plans can Improve RDP's Implementation

Whilst so far we have seen how RDP can improve rural landscape, particularly through AES, it is now time to see what Landscape Plans can do for RDP. As mentioned, landscape planning is a concept encompassing a variety of practices and tools. Contrary to rural policies and RDP implementation, which is easily comparable across EU Member States as it ensues from a common regulation, landscape plans may vary among countries in terms of contents, legal frames, responsible authorities. Although a thorough comparison of different situations is beyond the scope of this chapter, we can outline some common features of landscape plans and highlight the different rationality underpinning them, in comparison to RDP.

In general, landscape plans identify and classify areas based on their landscape features and define rules and orientations to prevent undesired modifications and promote the maintenance of valuable landscapes. They can also contain strategies and policy options to foster changes in currently "undesired" landscape: (e.g. peri-urban areas, brownfield, and intensive cropland) or set rules to mitigate the impact of artificial features on existing landscapes. Both RDP and landscape plans share an analytical apparatus that serves to classify different portion of the land. However, in the case of RDP, the rational underlying this is, again, sectorial: for the period 2007–2013, RDP had to classify the whole territory in four main categories: urban centres, intensive agricultural rural areas, intermediate rural areas, and marginal rural areas, based on the characteristics and relative importance of agricultural activity. Conversely, landscape plans identify landscape ambits based on a more holistic analysis, taking into account the ecological, cultural, historical and perceptive dimensions. In doing so, they feature a territorial approach: the kind and intensity of agricultural activity is one of the factors taken into account to delimit landscape ambits, but it is not the only one. On the other hand, they generally contain a set of rules varying in terms of prescriptive power and legal status (form binding regulations to directives, orientations, and simply moral suasion) but are not associated to a budget. They set the "rules of the game" (what can or cannot be done in certain areas) but do not have the resources and the capability to directly realize actions that can modify the landscape. Their implementation is therefore mostly indirect and transferred to other tiers of the decision-making and planning hierarchy or used as reference for other sectorial plans or programs. In that, their approach may be considered top-down, even if the plan's elaboration itself may have occurred using a participatory, inclusive approach.

Furthermore, whilst landscape plans may establish certain rules to prevent land abandonment or conversion of agricultural land to other land uses, they have no or very limited leeway to affect the type of management on agricultural areas, as generally this kinds of regulations fall under the responsibility of sectoral agricultural laws and policies (see also Cassatella and Seardo, *infra*, Chap. 3, Sect. 5.3).

To summarise, landscape plans are characterised by an holistic, territorial approach, which is the premise to design effective landscape preservation policies, but do not directly translate them into concrete, proactive measures, and delegate their implementation to other government tiers and sectorial plans/programs, often with a top-down, hierarchical approach. On the other hand, RDP feature a more sectorial and less holistic analytical apparatus, which is not geared to grasp the complexity of the interrelations that form the landscape, but have the capacity to implement actions (through funding) that directly affect landscape. Their overall effects is, however, the result of a large number of individual decisions made by farmers, acting according to an economic rationale, weakly influenced by a territorial one. The challenge is therefore to shift the mode of governance of RDP towards a territorial governance approach. In the next section, a case study is presented to provide concrete options in this sense.

2.4 Case study: Pursuing Synergies Between the Territorial Plan of the Turin Province and the Piedmont RDP

The case study presented seeks to show how joint implementation of Landscape/spatial plans and RDP can deliver mutual benefits towards the achievement of their objectives. The instruments examined are the RDP of the Piedmont Region (NUTS 2), Italy, and the Provincial Territorial Plan (PTP) of the Province of Turin, one of Piedmont's 8 Provinces (NUTS 3). Turin's Province is one of Italy's largest, with a total area of 6,827 square km and a population of 2.3 millions. Its territory presents a variety of landscapes, comprising mountains areas in the west (classified as marginal rural areas by the regional RDP), hilly areas in the central-eastern part, intensively cropped land in the central part, as well as the metropolitan areas of Turin, accounting for about 1.7 millions inhabitants (Fig. 2.1). The total Utilized

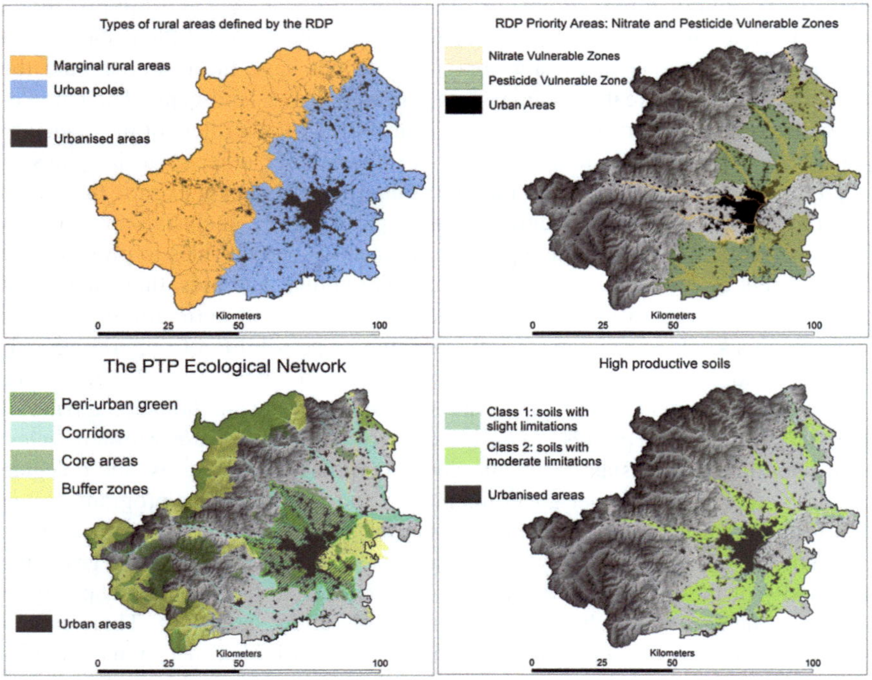

Fig. 2.1 Territorial classifications of the RDP and the PTP used to determine the priority areas for targeting the examined AES. From *left*-above clockwise types of rural areas; nitrate and pesticide vulnerable zone; ecological network; highly productive soils

Agricultural Area as of 2010 is 227,165 ha, of which 34 % is devoted to maize and wheat; 50 % to pastures and grasslands; 6.3 % to harvested woods (ISTAT 2013).

The new PTP entered into force in July 2011. It sets rules, regulations, directives and orientations for spatial development to be implemented by local land use plans. Furthermore, it contains a detailed analysis of the different types of landscapes in the provincial territory and their evolution over time, as well as policies and rules for landscape preservation. The plan pursues a strong limitation of new urbanization and soil sealing by requiring that municipalities subdivide their territory in three types of areas:

- Dense urban areas—where the urban fabric is consolidated, and where the main urban services and facilities are located;
- Transition Areas—localized at the margins of dense urban areas and characterized by a less dense urban fabric;
- Free Areas—where agriculture or natural assets are the predominant land uses.

Following this classification, the PTP establishes that no new developments are allowed in free areas. This policy is relevant for rural areas, which constitute the majority of the free areas. To determine the potential synergies with the Regional

RDP, first the specific PTP's policies and rules for agricultural areas and forestry were identified, then they were matched with measures of the RDP pursuing similar objectives. The result of this exercise is summarized in the Table 2.1. In the first column, reference to the PTP's regulations and their legal binding power are provided[1]; in the second column the PTP policies are summarised, whilst in the third one the RDP measures that can be related to the latter are identified, both with reference to current (2007–2013) and forthcoming (2014–2020) RDP.

The table shows that the PTP and he RDP share common objectives, and that each PTP's policy concerning agricultural areas and forestry can be related to one or more measure of the RDP (both the current and the future one). In particular, the PTP pursues two main objectives as regards agricultural areas: the first one is to support and maintain agricultural activity in highly productive soils, namely soils classified in classes 1 and 2 according their land capability.[2] Here the focus is on the productive capacity of such areas rather than on their ecological value, although by contrasting land use changes the plan also preserves them from urbanization. The second objective concerns peri-urban agricultural areas, i.e. remnants of free areas at the urban fringe, characterized by low landscape attractiveness, high fragmentation and low naturalness. Here the focus is on the ecological value of this sites and their importance in contrasting environmental negative effects from adjacent urban areas. Interestingly enough, the PTP explicitly states that to foster agricultural activity in these sites, synergies with the objective of the CAP are to be pursued.

The PTP also define the Provincial Ecological Network, articulated in the following main elements:

- **Core areas**: protected areas and Natura 2000 sites;
- **Ecological corridors**: buffer strips along rivers and main creeks;
- **Buffer zones**: other areas with significant environmental and *landscape* features; they comprise national and regional landscape designated areas and other sites with ecological and cultural value.

The PTP establishes a set of policies for the preservation and enhancement of the Ecological Networks, as summarised in the last row of Table 2.1. As regards rural areas belonging to the network, the PTP policy is to promote the presence of natural and semi-natural habitats, characterized by autochthonous species with ecological

[1] Regulations of the PTP are articulated in: (i) *prescriptions*: legally binding laws that have direct efficacy or that must be implemented by lower government levels; (ii) *directives*: rules to which municipalities and other bodies must comply or implement through their acts, unless a sound and reasoned justification is provided; (iii) *orientations*: more general policies that shall be implemented by municipalities and other bodies, with a less stringent binding power.

[2] The land capability system was developed by the US Department of Agriculture. Soils are grouped primarily on the basis of their capability to produce common cultivated crops and pasture plants without deteriorating over a long period of time. Overall, 8 classes are identified: Class 1 soils have slight limitations that restrict their use, whilst Class 2 soils have moderate limitations that restrict the choice of plants or that require moderate conservation practices.

Table 2.1 Correspondences and synergies between policies/regulations of Turin's PTP, the current regional RDP and the priorities and focus areas of the new RDP regulation

Reference in PTP	PTP Policies	RDP measures/priorities	
		2007–2013 RDP measure	2014–2020 priorities and focus ares
Art. 17.8 (prescription to be implemented by land use plans)	Good and excellent agricultural land shall be preserved. No new developments (either residential or productive) are allowed on these areas unless strictly necessary and no other alternative is feasible. Agricultural activity on these areas shall be maintained and encouraged	All measures of Axis 1 aimed at improving the competitiveness of the agricultural sector, in particular Measure 121 (farm modernisation) Measure 123: Adding value to agricultural and forestry products Measure 124: Cooperation for development of new products, processes and technologies	Priority (2) Enhancing farm viability and competitiveness and promoting innovative farm technologies
Art. 26 Forestry sector	Prescriptions: New developments in woodlands are not allowed; low density trees areas in the plain are to be preserved; if woodland clearance is necessary, it must be offset by an equivalent area through afforestation	Measure of Axis 2 related to forestry, in particular Measures 221 first afforestation of agricultural land; 222 first establishment of agro-forestry systems on agricultural land; 223 first afforestation of non-agricultural land; 225 forest-environment payments; 226 restoring forestry potential and introducing prevention actions; 227 Non-productive investments	Priority (4) Restoring, preserving and enhancing ecosystems related to agriculture and forestry

(continued)

Table 2.1 (continued)

Reference in PTP	PTP Policies	RDP measures/priorities	
		2007–2013 RDP measure	2014–2020 priorities and focus ares
27 Highly productive agricultural areas (directive)	27.2 Any transformation on such areas shall not prejudice their agricultural function	As above	As above
28 specialised and irrigated crops	Specialised cultivations (DOC or IGP products) shall be preserved and no other land use except agriculture is allowed on these areas, unless this is strongly motivated and no feasible alternatives are possible	Measure 124 Measures 132 and 133: Participation in food quality schemes	As above +(3) Promoting food chain organisation, including processing and marketing of agricultural products, with a focus on: improving competitiveness of producers by integrating them into the agri-food chain through quality schemes, adding value to agricultural products, promotion in local markets and short supply circuits, producer groups and inter-branch organisations
34 Peri-urban areas (directive)	• Enhance their ecological value by improving their vegetal equipment; • Improve their visual appearance by clearly demarcated the urban-rural edge	Measure of Axis 2 aimed at enhancing landscape features, in particular Measure 216 “Non-productive investments” that support habitat restoration, creation of landscape features and structures needed to ensure coexistence of farming activities and wildlife	Priority (4) Restoring, preserving and enhancing ecosystems related to agriculture, with a focus on restoring, preserving and enhancing biodiversity, as well as the state of European landscapes
	• Incentivize the permanence of agriculture activity • Promote multifunctional uses according to the principles of the CAP	Measures of Axis 3 aimed to support diversification, in particular: Measure 311 “Diversification into non-agricultural activities” and Measure 312 “Support for business creation and development”	Priority (6) Promoting economic development in rural areas, with a focus on facilitating diversification, creation and development of small enterprises and job

(continued)

Table 2.1 (continued)

Reference in PTP	PTP Policies	RDP measures/priorities	
		2007–2013 RDP measure	2014–2020 priorities and focus ares
35 Ecological network	Orientations: preserve biodiversity, preserve remnants natural and semi-natural areas in the Po Plain; promote the presence of semi-natural vegetation in rural areas Promote the creation of the ecological network also through horizontal and vertical coordination Promote landscape improvement through the realization of soft mobility (cycle routes, trails), landscape features (hedgerows, trees etc.) and cultural features	Measure of Axis 2 aimed at enhancing landscape features, in particular Measure 216 "Non-productive investments" and 214 "Agri-environmental payments"	Priority (4) Restoring, preserving and enhancing ecosystems related to agriculture, with a focus on restoring, preserving and enhancing biodiversity, improving water management, as well as the state of European landscapes

functionality, which can be clearly implemented through AES aimed at creation and maintenance of semi-natural elements on farms. Another objective of the Ecological Network is to promote ecological and landscape restoration through mitigations and compensations such as strip areas, hedgerows, tree lines and so on, which, again, is a typical feature of AES. To achieve these objectives, it is stated that coordination between different levels of government and planning, as well as with other policy sectors, is needed, thus implicitly fostering inter-sectoral coordination between, inter alia, the agricultural sector and rural development policy design. It clearly emerges how RDP's measures, in particular AES, would represent an effective way to implement the PTP's policy concerning rural areas and the ecological network. On the other hand, the PTP provides a detailed analysis of the provincial territory's landscape and ecological features that in turn can be useful for the RDP, particularly to identify priority areas on which to implement AES.

To this end, in the second part of the exercise presented here, a spatial analysis was performed using GIS to identify and rank priority areas to target AES by refining the priority system used by the RDP with the more detailed analysis provided by the PTP. As said earlier, the RDP establish scores to identify priority areas for AES implementation. Here we examine two AES, corresponding to measures 214.1 (integrated farming) and Measure 216 (non-productive investments) of Piedmont's RDP in 2007–2013 (but the same measure apply to all RDP in Europe). They are described in Box 1. Very similar measures will be present in the 2014–2020 period under priority 4.

Priority areas were identified for these measures by the RDP based on the main type of agricultural area defined above (intensive, intermediate, marginal areas, urban poles) and their environmental values: protected areas, Natura 2000 sites,

Box 1: AES examined in the case study

Measure 214.1—Application of integrated production techniques

This measure is one of the main AESs of Piedmont's RDP, requiring farmers to adopt the norms of *integrated production* established by the Regional Authority, which impose low input farming practices by establishing restrictions on the quantity and quality of fertilizers and pesticides allowed. Crop rotation is required for non-permanent crops. This AES also envisages the realization of permanent grass covers, turfs, biologic mulching and artificial nests for birds, but as optional actions.

Measure 216—Non Productive investments

Aim at preserving the traditional rural landscape and supporting biodiversity by financing respectively the maintenance and creation of natural and semi-natural elements and landscape features on agricultural land, such as: tree planting/management, restoration of wetlands and moorlands, hedgerows, terraces, grass covers and strips and set-aside areas, artificial nests for birds and bats.

Nitrate and pesticide Vulnerable Zones. In designing the annual call for applications, different scores were given to rank and select applications based on farms' localisation. The scores are additive, as a parcel may belong at the same time to more than one priority area. Such scores thus reflect the spatial targeting policy of the Managing Authority, and the relative importance assigned to different sites based on the expected environmental benefits of these AES. The score are shown in the second and third column of Table 2.2, whilst the different types of priority areas are shown in Fig. 2.1.

As part of this exercise, an alternative set of scores was derived based on the analysis of the PTP policies on rural areas and the ecological network. These scores (shown in the fourth and fifth column of Table 2.2) were assigned taking as reference those defined by the RDP, and subsequently refining them to integrate PTP's analyses and policies: for instance, Natura 2000 and protected area are the core areas of the Ecological Network defined by the PTP and are also priority areas for the RDP with a score of 10 (the highest one), so this value was maintained. Buffer zones are another element of the ecological network, but are not present in the RDP: in this case, a score of 7 was given, reflecting the fact that they are areas with high ecological value (identified by the PTP), but less important than core areas. Another main difference is the scores given to areas based on their agricultural characteristics. The RDP assigns priority to urban poles and intensive agricultural areas, thus giving priority to the function of AES in mitigating environmental negative effects produced by intensive farming practice. However, in the classification of the RDP, the Province of Turin only comprises marginal areas (mountain territories in the west part) and urban poles (the rest of the province, Fig. 2.1). This classification derives from the one used by the European Commission to identify predominantly rural and urban area (see Rega, *infra*, Chap. 1), mainly based on population density, but does not adequately reflect the territorial feature of the Turin's province where actually high productive agricultural areas and intermediate ones could be identified. Conversely, the PTP provides a more accurate classification, particularly by identifying peri-urban areas more in detail. In determining the second set of scores, this classification was used to assign priority to high productive soils, where more intensive farming practice is located, for measure 214.1 (integrated farming), and to peri-urban areas for measure 216 (the PTP policy is in fact to enhance their ecological value).

These two sets of scores—reflecting the policies of the RDP and the PTP respectively—applied to two different AES, were then used to produce four different maps displaying the different degrees of priority for each examined AES. Results are shown in Fig. 2.2 and discussed in Section 2.5.

Table 2.2 Scores assigned to priority areas for the examined AES by the current Piedmont's RDP and new scores assigned in the exercise as result of the consideration of the PTP analyses and policies

Priority areas	Scores assigned by Piedmont RDP		New scores assigned considering the PTP policies	
	214.1	216	214.1	216
No priority areas	2	0	2	0
Nitrate vulnerable zones	6	6	6	6
Pesticide vulnerable zones	8	8	8	8
Natura 2000 sites/protected areas	10	10	10	10
Buffer strips along main rivers/ecological corridors	0 (not present)	10	10	10
Areas with significant environmental and landscape features (buffer zones of the provincial ecological network)	0 (not present)	0 (not present)	7	7
Urban poles	4	4	0 (not present)	0 (not present)
Intermediate areas	3 (not present in the province of Turin)	3 (not present in the province of Turin)	0 (not present)	0 (not present)
Intensive agricultural areas	5 (not present in the province of Turin)	5 (not present in the province of Turin)	0 (not present)	0 (not present)
Marginal areas	1	1	1	1
Soils with land capability class 1 or 2	0 (not present)	0 (not present)	8	3
Peri-urban areas	0 (not present)	0 (not present)	4	8

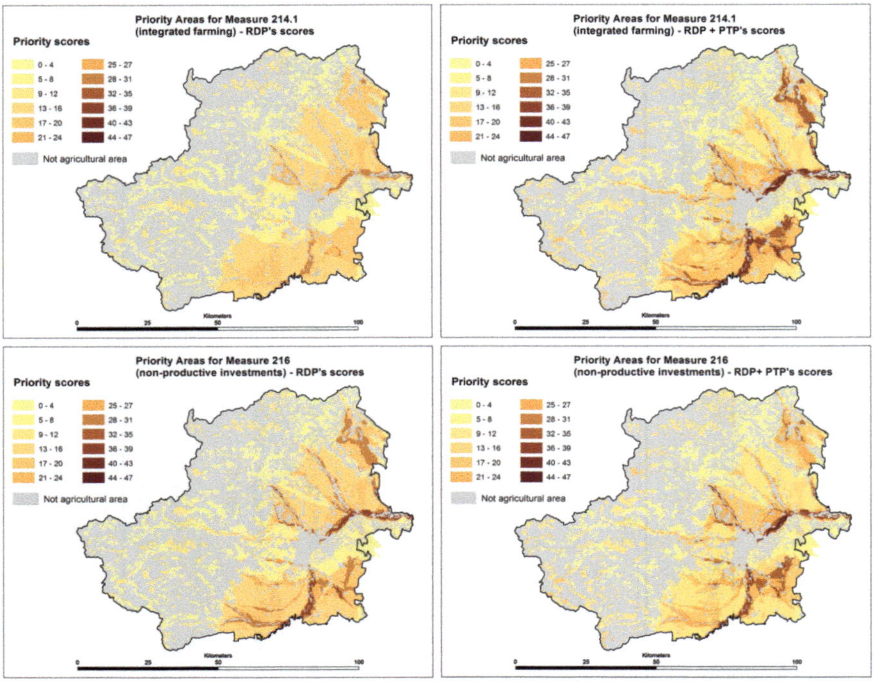

Fig. 2.2 Identification of priority areas for integrated farming (*above*) and non-productive investments (*below*) according to the RDP's scores (*left*) and the set of scores as integrated by the PTP (*right*)

2.5 Discussion: Towards a Landscape and Territorial Governance Approach to Agri-Environmental Schemes

Figure 2.2 shows that integrating the spatial priorities of the RDP with information and policies of the PTP allows to produce a more detailed and articulated classification of the provincial territory, based on analysis comprising its landscape dimension. At the same time, the RDP represents an effective mean to implement PTP's landscape preservation policy, as AES offer an ideal platform to realize the kind of concrete actions the plan envisages: creation and maintenance of natural and semi-natural habitats in rural areas and mitigation of harmful effects from intensively cropped areas. Moreover, financially supporting farmers with RDP measures concurs to the overall objective of maintaining agricultural activity and preserve agricultural land from urbanization, which is a key objective of the PTP. Figure 2.2 in turn provides RDP managers with more detailed information about priority areas for AES implementation. Future calls for application could therefore be designed by taking into account the specific territorial features of eligible areas, and more detailed scores used to rank received applications. This exercise was limited to the province of Turin, but similar analysis could be easily extended to the whole Regional territory by taking into account the analysis provided by the forthcoming Regional Landscape Plan, once it comes into force.[3]

Beyond more articulated scores to priority areas, results shown in Fig. 2.2 can be used to foster what we referred to as a "territorial governance approach" to AES and RDP in Sect. 2.3. The concept of governance has been the subject of a large theoretical debate over the last two decades (Stead 2013; Davoudi et al. 2008); it is beyond the objectives of this chapter to further elaborate on it: suffice here to recall that "government refers to the dominance of State power organised through formal and hierarchical public sector agencies and bureaucratic procedures, while governance refers to the emergence of overlapping and complex relationships, involving 'new actors' external to the political arena" (Painter and Goodwin 1995).

The main elements of governance may be summarised as follows (Davoudi et al. 2008):

- Vertical coordination between actors and policies;
- Horizontal coordination among actors and policies;
- Participation and involvement of civil society and organised interests.

The term "territorial governance" has gained momentum recently in the EU research and policy agenda, although it is not always clear how it differs from "plain" governance (Stead 2013). Davoudi et al. 2008) distinguish the specific features of territorial governance based on two conceptualization of the "territory" (i) as a "social and political construction" (a concept they derive from Bagnasco and Le Galès 2000); (ii) as "territorial capital".

In the first case the focus is on cooperation between different agents and collective action that is "the actions undertaken by a set of actors that are related to the solution of a collective problem" (ibid, p. 35). In this frame, "territorial governance is an organisational mode of territorial collective action, based on openness and transparency of the process itself, on cooperation/coordination among actors (horizontally and vertically), and in a framework of a more or less explicit subsidiarity" (ibid.).

In the second case, the focus is on the place-specific and path-dependent elements to be found in different regions, comprising natural features, heritage goods, common goods, social knowledge, and institutional capacity. By integrating these two conceptualizations, the authors define territorial governance "as the process of organization and coordination of actors to develop territorial capital in a

[3] The Regional Landscape Pan of Piedmont Region was adopted in 2009. It contains detailed analysis of Piedmont's landscapes and regulations and directives applying to them. Similarly to the PTP, it also define a Regional Ecological network. In this case study, the PTP was consider instead of the Landscape Pan because the latter has not been officially approved yet, so it's not into force as of February 2014. The PTP instead is into force and municipalities are called to amend their local land use plans accordingly.

non-destructive way in order to improve territorial cohesion at different levels" (ibid, p. 37). They therefore add a fourth element specific of "territorial" governance to the three ones mentioned above (that apply to "plain governance" as well), namely "Territorialized actions", i.e. actions that are not simply localized in a certain area, but that are based on the shared valorisation of local specificities.

Starting from it, Stead (2013) further elaborates on the conceptualization of "territorial" governance and identifies three main characteristics that distinguish it from "plain" governance:

- the process of managing territorial dynamics (i.e. governance that is specifically focused on managing territorial development);
- the monitoring and assessment of territorial impacts;
- the process of delineating boundaries for dealing with different policy questions or problems.

The case study presented in Sect. 2.4, and more in general the integration between landscape/spatial plans and RDP, represent a paradigmatic example of how the territorial governance concept could be made operational. By integrating a spatial plan elaborated at the Provincial level with a regional program, the case (potentially) represent in fact an example of both horizontal and vertical coordination (Region-Province; Spatial/landscape Planning-Agriculture); by influencing the spatial development of rural areas it is a process of managing territorial dynamics and development; it produces effects that must be monitored and assessed, and allow to define boundaries with "variable geometry" depending on the problem to be addressed (e.g. different priority areas for spatial targeting varying depending on the considered AES and policy objectives, as depicted in Fig. 2.2).

Referring to the concept of "territorialized actions" presented above, it can be argued that the identification of priority areas as result of the exercise illustrate in Sect. 2.4 allows not just identifying "sites" on which to localize the actions, but "territories", comprising natural and landscape features as well as local actors and institutions. This approach would allow extending the territorial governance approach of some measures of RDP (local development of rural communities, former Axes 3 and 4, now priority 6) to AES as well.

There are a number of actions and innovations that would concur to AES's territorialisation: first, the identification of specific territories would allow to define more detailed objectives and actions to be implemented, depending on their distinguishing characteristics and needs. Referring to the Turin's PTP, in peri-urban areas priority would be given to the realization of semi-natural elements and landscape features that both improve the visual appearance of such areas and their ecological values, as hedgerows, tree lines, ponds, or wet areas. In highly intensive areas, action instead would focus on reduction of chemical inputs. In buffer zones of the ecological network, actions would depend on the specific characteristics that make the site valuable: in some cases the increase of the semi-natural vegetation could not be a desirable option, if this decreases the visibility from particular vantage points (see Cassatella and Seardo, *infra*, Chap. 3 for an example). In other cases the priority could be to support biodiversity, thus the installation of artificial nests or the provision of food through no harvested areas would be the first option.

The selection of specific territories would also be a way to identify potential stakeholders or institutions that could act as catalysts for joint management of AES. In the case of protected areas or Natura 2000 sites, the public authorities in charge of their management could act as single points to collect application from farmers whose land is located on such areas. This would ease the administrative burden as there would be a single, collective application to be managed instead of a multitude of individual applications. This in turn might decrease the transaction costs for single farmers, thus increasing the total uptake of the AES, and contribute reaching the minimal threshold that often is required for an AES to be effective (see Sect. 2.2). Agricultural property is often fragmented in protected areas and Natura 2000 sites, which often make applications from individual farmers not eligible, as they do not reach the minimum area required to enrol; in the case of collective applications, this problem would be overcome.

Often, specific territories not only feature distinguishing natural/landscape characteristics, but are also similar cultivations and products: aggregation of stakeholder based on territorial clusters could therefore also allow to tailor made AES according to the cultivated crops (e.g. desirable AES would vary significantly if applied to permanent crops like orchards or vineyards vs arable crops). In turn, this could also be a way to promote and foster food chain organizations and develop "common" environmental practice that could trigger emulation among producers in the same branch. Public-Private partnership can be established to this end, comprising for example municipalities, Provinces, managing bodies of protected areas and farmer's association, environmental NGOs or consumers associations.

France provides a good example of such an approach to AES: the national RDP envisages the so called *Mesures agro-environnementaux territorialisés* (territorialised agri-environmental schemes) whereby for each measure a local "leader" is identified (it may be a public body, as well as a private one) in charge of establishing a partnership and implement the AES according to the specificities of the identify territory. The managing authority issues a public call, so that potential beneficiaries are encouraged to develop integrated projects and strongly commit to them. In this way, the local knowledge is mobilised and more efficient implementation is expected (Jeanneaux et al. 2011).

The proposed shift toward a territorial governance in RDP and AES put forward here resonates with recent works by scholars from different research fields, although not necessarily the authors explicitly refer to "territorial governance". A part the cited work of Jeanneaux et al. (2011) regarding the French context, Etxano (2012) for instance argues that integration of Protected Areas assessment and management and RDP would lead to more effective outcomes in terms of socio-economic and environmental benefits, by including local communities and pursuing synergies between RDP policies and protected areas' conservation objectives. The argument is supported by the presentation of a case study in the Basque Country, Spain. Dulcire et al. (2012) explicitly refer to "territorial governance" in

RDP by presenting a case study on the application of the "Sustainable Agricultural Contract" under the French law on Agriculture as a way to promote "a new mode of governance in which the state liaises with local stakeholders to identify and implement actions" in the agri-environmental sector (ibid, p. 71). A case study from the island of Réunion (French Overseas territory in the Indian Ocean) shows that joint, territorialised agri-environmental projects can be realized if the effective participation of stakeholders at every stage is granted. In a recent study in the field of ecology, McKenzie et al. (2013) argue that joined-up, landscape-scale AES—i.e. single environmental management agreements for multiple farm units[4]—are likely to benefit a small but key group of species more than current 'farm-scale' schemes, while not disadvantaging species operating at smaller scales. They are also likely to provide additional co-benefits in relation to some Ecosystem Services. Result based on semi-structured interviews with farmers also indicate that the majority of them would be in principle favourable to collaborative AES, as they perceive that these have the potential to deliver better environmental benefits while using less farm resources than current AES.

2.6 Conclusions

This chapter addressed the issue of integration between landscape/spatial planning and Rural Development Programmes (RDP), with a focus on Agri-Environmental Schemes (AES). First, an overview off current (2007–2013) and forthcoming (2014–2020) RDP objectives and contents was provided. Subsequently, shortcomings in AES implementation were identified and discussed, in particular the lack of spatial targeting and coordination between beneficiaries, which makes a strong case for the argued integration with spatial/landscape plans. On a theoretic level, a distinction between a sectoral and a territorial rationale was made. The first strongly permeates current RDP and AES design and implementation, conceiving agriculture mainly as an economic activity. The territorial dimension instead conceptualises agriculture as one of the main activity shaping the territory and the landscape, and stresses its interlinks with landscape and environmental preservation. It was also shown that landscape plans generally feature a more detailed analytical apparatus and set rules and visions for landscape management with a holistic approach. On the other hand, their implementation is mainly deferred to other tiers of the government systems and they cannot realize direct actions. Conversely, RDP directly affect landscape through funding concrete projects, but their overall effects is limited by a narrow sectoral focus, and the lack of coordination between single agents undertaking these actions (farmers). A case study

[4] Examples of collaborative AES include the creation of networks of ponds and wet areas, targeted tree planting, coordinating the timing of harvest with neighbours, Creating/extending networks of hedgerows.

concerning the Piedmont RDP and the Turin Provincial Territorial Plan (PTP) was presented: synergies and convergences between the policies and objectives of these two instruments were identified, and the territorial/landscape classification provided by the PTP was used to refine the identification of priority areas on which to target two major AES.

The argument put forward is that this kind of exercises can foster a shift towards a *territorial governance* of AES, whereby not only actions are targeted to specific areas according to their landscape and environmental features, but joint projects based on the identification of local stakeholders and partnerships can be elaborated so to mobilise the local knowledge and capacities, with the overall aim of increasing their effectiveness. This entails increased horizontal and vertical coordination between public bodies and policy sectors, as well as increased involvement of civil society organizations on a territorial base. This proposal resonates with similar ones coming from different research fields, and intends to contribute to the forthcoming implementation of the new European Rural Development Regulation, as a way to pursue the implementation of the European landscape Convention and the EU objectives of sustainable development.

References

Bagnasco A, Le Galès P (2000) Introduction. European cities: local societies and collective actors?". In: Bagnasco A, Le Gales P (eds) Cities in contemporary Europe. Cambridge University Press, Cambridge, pp 1–32

CoE (Council of Europe) (2000) European landscape convention, Florence

Davoudi, S, Evans E, Governa F, Santangelo M (2008) Territorial governance in the making. Approaches, methodologies, practices. Boletin de la A.G.E. 46:33–52

Dulcire M, Chia E, Piraux M (2012) The multi-functionality of agriculture and territorial governance—a learning process in an Island environment (Réunion). In: Adisa RS (ed) Rural development—contemporary issues and practices. InTech, Rijeka, pp 55–74

Dupraz P, Latouche K, Turpin N (2009) Threshold effect and co-ordination of agri-environmental efforts. J Environ Plann Manage 52(5):613–630. doi:10.1080/09640560902958164

EC (European Commission) (2010) The CAP Towards 2020: meeting the food, natural resources and territorial challenges of the future. Communication from the commission to the European parliament, the council, the European economic and social committee and the committee of the regions. COM (2010) 672 final

ECA (European Court of Auditors) (2011) Is agri-environment support well designed and managed? Special report N° 7. Publications office of the European union, Luxembourg. http://eca.europa.eu/portal/pls/portal/docs/1/8760788. Accessed 12 Feb 2014

Etxano I (2012) New insights into the assessment of protected areas—integrating Rural Development. In: Adisa RS (ed) Rural development—contemporary issues and practices. InTech, Rijeka, pp 75–98

EU (European Union) (2013) Regulation (EU) No 1305/2013 of the European parliament and of the council of 17 December 2013 on support for rural development by the European agricultural fund for rural development (EAFRD) and repealing council regulation (EC) No 1698/2005. Official Journal of the European Union L 347/487. http://eur-lex.europa.eu/LexUriServ/LexUriServ.do?uri=OJ:L:2013:347:0487:0548:EN:PDF. Accessed 12 Feb 2013

Garrod G (2009) Greening the CAP: how the improved design and implementation of agri-environment schemes can enhance the delivery of environmental benefits. J Environ Plann Manage 52(5):571–574. doi:10.1080/09640560902958099
ISTAT (Istituto Nazionale di Statistica) (2013), 6° censimento generale dell'agricoltura in Piemonte. http://www.istat.it/it/files/2013/02/Agricoltura-Piemonte.pdf?title=Censimento+agricoltura+in+Piemonte+-+01%2Ffeb%2F2013+-+Testo+integrale.pdf. Accessed 14 Feb 2014
Jeanneaux P, Aznar O, Déprés C (2011) Les services environnementaux fournis par l'agriculture et leurs modes de gouvernance: Un cadre d'analyse économique. Reg Cohesion 1(3):117–144
Lefebvre M, Espinosa M et al. (2012) The influence of the common agricultural policy on agricultural landscapes. J. S. a. P. Report, European Commission, Joint Research Center
Lennox GD, Armsworth PR (2011) Suitability of short or long conservation contracts under ecological and socio-economic uncertainty. Ecol Model 222:2856–2866
Kleijn D, Sutherland WJ (2003) How effective are European agri-environment schemes on conserving and promoting biodiversity? J Appl Ecol 40:947–970
Mettepenningen E, Verspecht A, Van Huylenbroeck G (2009) Measuring private transaction costs of European agri-environmental schemes. J Environ Plann Manage 52:649–667
McKenzie AJ, Emery SB, Franks JR, Whittingham MJ (2013) FORUM: landscape-scale conservation: collaborative agri-environment schemes could benefit both biodiversity and ecosystem services, but will farmers be willing to participate? J Appl Ecol 50(5):1274–1280. doi:10.1111/1365-2664.12122
Painter J, Goodwin M (1995) Local governance and concrete research: investigating the uneven development of regulation. Econ Soc 24(3):334–356
Paracchini ML, Capitani C (2011) Implementation of a EU wide indicator for the rural-agrarian landscape. Joint research centre technical report EUR 25114 EN—2011. Publications Office of the European Union, Luxembourg. doi:10.2788/26827
Piorr A, Ungaro F, Ciancaglini A, Happe K, Sahrbacher A, Sattler C, Uthes S, Zander P (2009) Integrated assessment of future CAP policies: land use changes, spatial patterns and targeting. Environ Sci Policy 12:1112–1136
Rega C, Spaziante A (2013) Linking ecosystem services to agri-environmental schemes through SEA: a case study from Northern Italy. Environ Impact Assess Rev 40:47–53
Ruto E, Garrod GJ (2009) Investigating farmers' preferences for the design of agrienvironment schemes: a choice experiment approach. J Environ Plann Manage 52:631–647
Saraceno E (2002) Rural development policies and the second pillar of the common agricultural policy. Paper presented at the ARL/DATAR workshop on "Desirable evolution of the CAP: a contribution", Bruxelles, Belgium, 23 Sept 2002. https://www.tcd.ie/Economics/staff/amtthews/FoodPolicy/CourseMaterials/Readings/saraceno.pdf. Accessed 3 Feb 2014
Spaziante A, Rega C, Carbone M (2013) Enhancing SEA's effectiveness: spatial analysis of agri-environmental measures in the SEA of rural development programs. Scienze Regionali Ital J Reg Sci 12(2):93–116
Stead D (2013) The rise of territorial governance in European policy. Eur Plan Stud. doi:10.1080/09654313.2013.786684
Uthes S, Matzdorf B, Müller K, Kaechele H (2010) Spatial targeting of agri-environmental measures: cost-effectiveness and distributional consequences. Environ Manage 46(3):494–509
Vergamini D, Raggi M, Viaggi D (2013) Evaluating the potential contribution of spatially differentiated payments to the efficiency of agri-environmental measures: a resource allocation model for Emilia Romagna (Italy). Paper presented at the 2nd AIEAA conference "Between crisis and development: which role for the bio-economy", Parma, Italy, 6–7 June 2013
Whittingham MJ (2011) The future of agri-environment schemes: biodiversity gains and ecosystem service delivery? J Appl Ecol 48:509–513
Wunder S, Engel S, Pagiola S (2008) Taking stock: a comparative analysis of payments for environmental services programs in developed and developing countries. Ecol Econ 65:834–852

Chapter 3
In Search for Multifunctionality: The Contribution of Scenic Landscape Assessment

Claudia Cassatella and Bianca Maria Seardo

Abstract The attention recently paid to ecosystem services, which include cultural services, such as spiritual and aesthetic experiences, seems to encourage the consideration of landscape scenic values into rural development policies. However, existing theoretical frameworks doesn't clarify enough the differences between various landscape services, among which potential conflicts—deriving from multiple values related to the same spatial assets—may arise. A sound assessment of landscape services is necessary. The chapter aims to show that multifunctionality is a goal-oriented concept and an option, non to be considered an intrinsic character of landscape policies. Rural, environmental, landscape and spatial policies can partially share certain strategic objectives, spatial targets, and evaluation frameworks. The chapter illustrates existing techniques (as well as original proposals) for supporting landscape and rural policies through scenic landscape assessment, particularly: detailing categories of cultural services related to landscape amenity and developing scenic landscape indicators for environmental assessment frameworks. The second part of the chapter illustrates methodologies for the assessment of scenic landscape, their application in spatial planning and their potential application in rural policies, based on the Authors' research experiences on cultural landscapes in Italy. Evidence gained through the cases studies indicates that landscape scenic beauty can be protected and enhanced by integrating landscape and rural policies. The chapter drafts a theoretical framework and illustrates the practical outcomes by a wide range of possible planning measures.

The Chapter was jointly developed by both authors. However, Bianca Seardo wrote Sects. 3.1 and 3.4.2, Bianca Seardo and Claudia Cassatella Sect. 3.2, Claudia Cassatella Sects. 3.3–3.5.

C. Cassatella (✉) · B. M. Seardo
Interuniversity Department of Regional and Urban Studies and Planning (DIST),
Politecnico di Torino, Viale Mattioli 39, 10125 Torino, Italy
e-mail: claudia.cassatella@polito.it

B. M. Seardo
e-mail: bianca.seardo@polito.it

C. Rega (ed.), *Landscape Planning and Rural Development*,
SpringerBriefs in Geography, DOI: 10.1007/978-3-319-05759-0_3,

Keywords Cultural ecosystem services · Landscape multifunctionality · Rural landscape · Scenic landscape assessment · Landscape policies · Landscape planning

3.1 Landscape Amenity: A Politically Significant Issue

3.1.1 Should We Care for the Appearance of Rural Landscapes?

As regards perception, the idea of 'landscape' identifies a conceptual sphere that differs from that of 'environment' or 'territory', a sphere where image plays a fundamental role (Cassatella 2011, p. 105). It expresses the identity of a place, it facilitates our appreciation of its values and it is linked to the satisfaction felt by inhabitants when considering the quality of their surroundings.

The European Landscape Convention (CoE 2000) defines landscape as 'an area, as perceived by people' (Article 1) and recognises the fundamental role landscape plays in the 'formation of local cultures', 'a basic component of the European natural and cultural heritage, contributing to human well-being and consolidation of the European identity' (CoE 2000, Preamble).

Leisure and the enjoyment of the landscape have become significant factors in rural development, as emerges from the 'commodification' of certain rural landscape functions not directly involved in food production. The legibility of the landscape is a fundamental characteristic necessary for landscape orientation and use (Lynch 1960); turning former, intensively exploited landscapes into 'legible' ones is the principle behind the work done in many rural areas of historical and cultural value (Brinkhuijsen 2007).

The aesthetic qualities of particular rural landscapes help to make them an expression of the cultural identity of entire peoples. Italy protects the landscape thanks to its Constitution (1947); the Cultural Heritage and Landscape Code (the *Codice dei Beni Culturali e del Paesaggio*, Italian Republic 2004) 'safeguards landscape in relation to aspects and characters which constitute a material and evident representation of national identity' (Article 131).

At an international level, UNESCO lists and protects many rural landscapes as 'cultural landscapes' for—among other things—their singular legibility and scenic importance.[1] The internationally famous beauty of particular rural landscapes can become a promotional tool aiding local agricultural production, a factor in its

[1] In Italy alone we need only mention the rural landscape of the Val d'Orcia, often used 'to exemplify the beauty of well-managed Renaissance agricultural landscapes', and the Cinque Terre area, protected due to the 'exceptional scenic quality [of its coastal] intensively terraced landscape' (UNESCO 1998).

commercial success and consequently that of the local economy (as is the case with wine production in the Cinque Terre area, in Italy).

In 2012, the Italian Ministry of Agriculture, Food and Forestry Policies instituted the National Observatory of Rural Landscape, Agricultural Practices and Traditional Knowledge, responsible for surveying, research and the development of general principles and guidelines for the protection and enhancement of the rural landscape (Italian Republic, *Ministero delle Politiche Agricole* 2012; Agnoletti 2012).

If we look at the policies of other European countries, we find that the 'enjoyment of the countryside' is one of the objectives pursued by public policy in the UK (DEFRA 2000, p. 133). In the Netherlands, policies for rural territories work alongside strategies set up to safeguard landscape structure and the enhancement that is useful to more established economic and productive measures designed for the agricultural industry, thanks to national policy documents such as the *Belvedere Memorandum* (VROM 1999) and the *Agenda Landschap* (VROM and LNV 2008).

3.1.2 The Experience of Landscape Beauty as an Ecosystem Service

In theoretical frameworks of multifunctionality, landscape's aesthetic dimension is considered part of the cultural ecosystem services, the subject of increasing interest in the 'ecosystem services agenda' (Daniel et al. 2012) and in studies addressing the economic value of the landscape (Van der Heide and Heijman 2013). Nevertheless, when it comes to the systematisation of ecosystem services, cultural ecosystem services are a category that attracts less interest compared to those that are usually the subject, for example, of landscape ecology (Antrop 2007). Daniel et al. (2012) stress the fact that cultural ecosystem services are currently neither adequately defined nor taken into account within the broader ecosystem services framework.

By comparing the most common ecosystem services (Table 3.1), we can draw a few conclusions. The leading cases of systematisation list among cultural ecosystem services many of the factors that are also the basis of current expert techniques for landscape characterisation and valuation (Landscape Institute 2013; Swanwick 2002). Moreover, cultural services performed by ecosystems can be assimilated with those provided by the landscape. Indeed, landscape studies also highlight similar systematisations, for example the one proposed by Natural England (2009).

Landscape is, essentially, perception (CoE 2000, Article 1). The systematisations of cultural services show that landscape amenity is not limited to its aesthetic or visual dimension, but it also consists of other aspects: identity, spirituality, cultural diversity, a sense of the past, active enjoyment and its educational role

Table 3.1 Ecosystem cultural services and landscape services within some theoretical frameworks

	MEA (2005) Cultural services of ecosystems	TEEB (2010) Cultural and amenity services of ecosystems	NE (2009) Cultural services of landscape
Identity	Cultural landscapes and heritage values		A sense of history (or heritage)
	Cultural diversity and identity	Inspiration for culture, art and design	A sense of place (identity, home)
Aesthetics	Aesthetics	Aesthetic information	
	Inspirational services		Inspiration
	Spiritual services	Spiritual experience	Spiritual
			Escapism
			Calm (relaxation, tranquillity)
Recreation	Recreation and tourism	Recreation and tourism	Leisure and recreation
		Information for cognitive development	Learning (education)

Cultural services essentially coincide with landscape services. The order has been adapted in order to highlight this correspondence with regard to the aspects of identity, aesthetics and recreation

(Table 3.1). We can therefore assert that the beauty of a landscape is already encompassed by the functions still attributed to ecosystems. Nevertheless, we do come across issues that need to be addressed.

Reference to more specific services—and their identification—is weak or entirely lacking. As regards this aspect, systematisation does not seem to be adequately based on more recent progress made in specialist fields. For example, if more detailed services were highlighted, it would be easier to associate them with their respective sets of landscape indicators as developed in field literature to date (and as distinct from environmental ones, Cassatella 2011).

Our proposal is referring to three main, separate dimensions that can be used to analyse and characterise rural landscape amenity:

- a memorial function, i.e. the image that contributes to recognising and passing on identity (associated values: a sense of history, symbolic value, significance, distinctiveness, unicity, fame, integrity…);
- an aesthetic function, linked to contemplative enjoyment (associated values: beauty or scenic quality, tranquillity, sonority, variety, richness…);
- a recreational function, equivalent to its possible active use, i.e. its use in terms of outdoor activities, tourism, as well as its educational and informative use (usability).

The next section will examine the possible relations between these and other functions of the rural landscape.

3.2 Multifunctional Landscape from Theory to Practice: Trade-offs in the Landscape

Scientific attention to the paradigm of landscape multifunctionality was raised in the context of agricultural policies and their evaluation (Rentig et al. 2009). One of the earliest and best-known definitions of the concept is as follows (see also Rega, *infra*, Introduction): "multifunctionality refers to the fact that an economic activity may have multiple outputs and, by virtue of this, may contribute to several societal objectives at once" (OECD 2001, p. 11).

The question which the OECD poses for agriculture (2001, p. 9), "multifunctionality: a characteristic or an objective?"[2], also applies to landscape: Considering multifunctionality an intrinsic characteristic of landscape leads to an underestimation of the potential negative interactions (or "trade-offs") between different activities and policies, even if they all claim to be "sustainable".

The assumption that diverse environment and landscape functions can co-exist "at once" has major implications for policy making and evaluative frameworks: landscape functions are usually meant to be dependent on environmental ones. As a consequence, they are not adequately developed, or they are considered a priori equal in sign. This conception is quite clear in current models of environmental assessment (applied, for example, in Strategic Environmental Assessment and Environmental Impact Assessment), where landscape is a component of the environment, and is assessed by indicators derived from landscape ecology, or based on land-uses (Cassatella and Voghera 2011, p. 34). Even in Landscape Impact Assessment, in the field of perceptual indicators, the coincidence of naturalness and beauty is an implicit model. As a matter of fact, the most natural landscape may not always be the most scenic, however, most Visual Impact Assessment methods assume that they are (Churchward et al. 2013). The relationships between diversity of use/ecological diversity/visual diversity or between human-perceived naturalness/naturalness defined only by ecological characteristics needs to be clarified (Daniel 2001).

The consequence of the abovementioned paradigm is the idea that environment- and landscape-oriented actions can "naturally" coincide or otherwise derive mutual benefits. In practice, landscape values can conflict, depending on different social actors and perceptions. The following examples focus on the difficulty of managing the relationship between scenic values and other environmental functions.

The famous Dutch national ecological network project for the Randstad Holland envisaged the creation of 'new wilderness' areas with large-scale forestation

[2] "There are essentially two approaches to the analysis of multifunctionality. One is to interpret multifunctionality as a characteristic of an economic activity. (...) This view can be termed the "positive" concept of multifunctionality. The second way of interpreting multifunctionality is in terms of multiple roles assigned to agriculture. (...) This view can be termed the "normative" concept of multifunctionality" (OECD 2001, p. 9).

work carried out in rural areas. Some people noted that this also led to the destruction of traditional agricultural landscapes that had both memorial and identity-building value. In general, such 'greening' projects can create visual barriers to important sights or limit the openness of panoramic views. In some landscape enhancement projects, the clearing of vegetation is necessary to improve scenic aspects and encourage usability, to uncover historical landmarks or restore views of historical importance. In the Netherlands, this is the policy followed in some sites protected as Natural Landscapes, but such work can lead to negative repercussions for the environment. Yosemite National Park (California) launched its tree felling programme to restore the park's original views, but many people have raised concerns regarding the ecological effect of such felling.[3]

Moreover, many landscape evaluation models propose land consumption as an indicator. According to this indicator, any land taken away from the natural or agricultural habitat for urban use is a negative value. However, this implies that the anthropisation of landscape can never be a positive contribution! This generalisation is too extreme because it rules out the very concept of cultural landscape and it is an example of what happens when we borrow an environmental indicator and apply it to landscape evaluation models, supposing them to be the same thing. Urbanisation judged according to land consumption can obviously affect perceptive aspects such as the obstruction of views and light and sound pollution, but it can also produce new landscapes. Only the specific evaluation of landscape aspects can judge this (Cassatella et al. 2009).

In conclusion, multifunctionality is an option and not an intrinsic character of landscape. As a consequence, differences, potential positive and negative trade-offs between landscape function must be brought to light (Seardo 2012), in order to avoid conflicts and to maximize synergies between environmental, rural and landscape policies. Scenic landscape assessment can contribute to this effort by fostering the comprehension of perceptual aspects, through appropriate techniques. Methods for dealing with scenic landscape assessment are the subject of the next section. In Sect. 3.4, the contribution of scenic assessment to rural landscape policies and planning will be illustrated through case studies.

3.3 A Methodology for Assessing Scenic Landscape

Scenic landscape represents a challenge in research and in public policies (Cassatella 2014 in press) and a number of scientific methods (both qualitative and quantitative) for assessing scenic landscape are being proposed and experimented. This section illustrates a set of methods specifically developed and applied to Italian

[3] The tree felling programme—part of the Scenic Vista Management Plan (USDI-NPS 2010), an official planning document adopted by the park—works alongside a Strategic Environmental Assessment and was finally drafted in a participatory way in line with envisaged procedures.

cultural landscapes, and related to the field of landscape planning. Such methods derive from Landscape Character Assessment, but also include indicator-based evaluation.

The present section illustrates the methodological steps, while the next one will focus on the applications, pointing out the most relevant results in the perspective of landscape and rural policies.

Phase 1. Identification and (cartographic and iconographic) representation of scenic characteristics.

Firstly, landscape scenic features are identified by desk study, consisting of:

- sectoral analysis concerning the assessment of landscape attributes,
- listing designated (landscape, cultural or environmental) assets,
- other sources which can reflect social and identity value (literature, iconography, tourist guides),
- mapping and classification of potential observation points and routes.

Secondly, field surveys make it possible to verify the conditions of the resources identified above, the visual relations between them, the significance of the mapped vantage points and the existence of detrimental elements. The output is a map of scenic features and characters.

Phase 2. Visual analysis through Geographic Information Systems.

On the basis of a Digital Terrain Models (DTM) or a Digital Surface Models (DSM), a viewshed is calculated by a GIS space analyst and associated to each selected observation point or route.[4] The viewshed analysis depends on the choice of parameters related to the viewcone (radius, azimuth, width). In our applications, we used three different radii, relating, on the one hand, to perceptual phenomena (foreground, middle ground, background), and, on the other hand, to the possibility of applying different planning regulations to each one (the theory and methods are developed in Cassatella (2013a), a partial explanation in Cassatella, in press).

Using similar methods and parameters, the area of visual influence of emergent features, such as landmarks or detrimental elements, can be mapped.

Phase 3. Visual landscape indicators.

On the basis of the previous steps, a series of indicators can be calculated, using quantitative methods and GIS, or estimated, using qualitative methods (For landscape indicators of social and visual perception see Ode et al. (2008); Cassatella (2011)). In this chapter, as an example, we illustrate an indicator called "visual sensitivity" (VS).[5]

The indicator of visual sensitivity represents how much of an area is seen from the selected vantage points (i.e. the percentage of selected viewpoints which "see"

[4] The selection is made in phase 1.

[5] The term "landscape sensitivity" does not have a single, universally accepted definition. See, for example, Landscape Institute 2013 and USDA 1995.

an area), on a scale ranging from "highly visible" to "not visible", on the basis of the abovementioned viewshed analysis. Thus, an area which can be seen from all the selected viewpoints is classified as "highly visible" and, as a consequence, has a high VS. The numerical sensitivity value assigned using GIS is the sum of all viewsheds. It can, however, be transferred to another appropriate numerical scale.

Others indices can be obtained on the basis of VS, for example the percentage of landmarks which are visible, or not visible, which we can assume as an indicator of visual richness or "imageability"; the percentage of visual detriments which affect highly sensitive areas ("visual influence of detrimental elements").

All the outputs of the abovementioned phases (visual characteristics map, viewshed map, VS map, as well as photographic atlases, etc.) can be usefully implemented into landscape and spatial planning, in particular to help to foresee the effects of urban development plans on landscape, and, at a more detailed scale, to regulate and control building activity. Potential applications of scenic assessment are illustrated by case studies in the next section.

3.4 Scenic Landscape Assessment Applications for Policies and Plans: Case Studies in the Piedmont Region

The case studies presented here concern three Italian cultural landscapes, in the same region, representing different landscape situations in terms of the importance of agriculture and urban dynamics.[6] The Piedmont Region has a variety of rural areas, some of which are famous landscapes (such as the Langhe vineyards, or the rice fields of the Vercelli area). However, it is also strongly marked by urban development, so that rural and urban areas are inseparable, mainly in the perception of landscape.

The case studies call for different landscape strategies as they present different "rur-urban" dynamics: a protected rural landscape (Albugnano, surroundings of Vezzolano Abbey), a landscape in transition (San Martino Alfieri), and a peri-urban landscape (the metropolitan area of Turin). In 2009 the Piedmont Region adopted a Regional Landscape Plan (RLP), under which the three rural areas fall under different types of regulations, ranging from protective and prescriptive measures to general rules for "ordinary" landscape and a site-specific strategic project.

For each case study, three main aspects will be illustrated: the main scenic landscape characteristics and functions; (scenic) landscape quality objectives, proposed planning measures and tools.

---

[6] The contents of this section were originally developed based on the selection and summary of previous studies carried out by the Authors with several research groups. Full information and references in: Cassatella (2014) in press, Cassatella and Seardo (2012) and Cassatella (2013a, b).

3.4.1 A Rural Landscape Under Protection: Albugnano, Surroundings of Vezzolano Abbey

The first case study, Albugnano, is a nationally valued landscape, recognized by a designation act, in order to protect its historical and scenic value. The landscape is purely rural, alternating vineyards and woods, where historically relevant religious buildings constitute landmarks. Thus, the scenic characters are: openness and focality of the views, intervisibility between landmarks, and texture.

Amenity, tranquillity and spirituality, are the main landscape "services" which depend on scenic characters; therefore we can identify contemplation as the key function to protect and enhance. This means protecting panoramic views (not only the viewpoints, but also the area in the view), visual axes between viewpoints and landmarks, and conserving the integrity of texture, avoiding the intrusion of urban features.

Thanks to the act of designation, every urban development in the area is subject to a procedure of authorization, which assesses its "landscape compatibility". Moreover, in the framework of the RLP, each designated area is currently being associated to site-specific prescriptive regulations. Albugnano served as case study for developing and illustrating guidelines concerning scenic assets (Cassatella 2014 in press). Scenic assessment provided a map and an atlas of visual features and of visual relations (see Sect. 3.3, phase 1), and maps of the viewshed of observation points and panoramic routes (Fig. 3.1). Thanks to these maps, the authorities responsible for assessing the landscape compatibility of interventions can identify all the areas where whatever transformation may affect the scenery (Sect. 3.3, phase 2). The local authorities, which are responsible for local development plans, are supported in localizing interventions in areas with a low level of VS (Sect. 3.3, phase 3), from the planning phase onwards.

3.4.2 A Landscape in Transition: San Martino Alfieri

The hilly landscape of San Martino Alfieri is characterized by vineyards, natural woods, historic villages and farmhouses. The recent phenomena of urban sprawl, the success of urban models in the housing market, the abandonment of crops in areas which are too difficult to cultivate with modern techniques, and the construction of prefabricated concrete warehouses express the tendency towards a new coexistence between urban and rural elements of landscape. In contrast, the stated aim of planning documents is the conservation of the historic identity, and of its legibility, which depends on the relations between a number of traditional elements: field texture, types, scale, colours and materials of the settlements (in a word: landscape characters), hierarchy between symbolic landmarks and ordinary buildings.

Fig. 3.1 (*Above*) In-depth views are a distinctive character of the Albugnano nationally valued landscape. (*Below*) Landscape sensitivity mapped from observation points and panoramic routes

Table 3.2 Checklist for scenic character assessment of rural landscapes in San Martino Alfieri, Piedmont Region

Landmarks and visual relationships	Landmarks (near or in the distance), landmarks or frames outside the study area (summits, ridges, symbolic buildings, other settlements), axial lines, urban doorways, urban doorways aligned with landmarks, chapels right in front of the entrances of settlements, visibility of the skylines of the settlements from ordinary roads, inter-visibility among the settlements
Viewing places	Vantage points, vantage points with facilities, urban squares with vantage points, equipped and signaled scenic routes, equipped and signaled routes of environmental interest, routes along ridges, hillside routes and routes overpassing valleys
Background and frames	In-depth panoramas, scenic enclosures, hilly skylines, morphological features distinguishing watersheds, panoramas of the Alps
Visual detriments	Railways, power lines, quarries, transport infrastructure, settlements characterized by heterogeneity sited along ridges, panoramas on degraded landscapes: productive or commercial areas, urban sprawl within rural landscape
Identity elements	Places of collectively recognized value, other places or artefacts of social value
Materials, colours, textures	Colour of the topsoil, traditional building materials, floorings, stones used for walls and buildings
Areal characteristics	Signs of nature: water, woods; signs of traditional agriculture; signs of intensive agriculture: arable lands, specialized vineyards

Source Cassatella and Seardo (2012)

A new equilibrium between old and new characteristics must be achieved by the ordinary forms of land management and planning. The Landscape Character Assessment contributes with a detailed check-list of scenic characteristics related to the traditional landscape (Table 3.2), which may be used as a reference for protection measures, as well as for local design codes, building regulations, guidelines and for the assessment of new interventions. Local codes may also deal with types of vegetation and species to be preferred or avoided in green spaces: indeed, the inhabitants of this landscape tend to follow urban models in their gardens and public spaces too, abandoning traditional local elements and favouring cosmopolitan species.

The landscape analyses for San Martino Alfieri included studies on agronomy, landscape ecology, history and scenery, carried out by different experts. The overlapping of these analyses, in search for a synthetic assessment, revealed multiple and sometimes diverging values assigned to the same elements. The observed trade-offs between different landscape functions can be illustrated by the following example. The majority of routes in the area have a historic character, offer panoramic views and, thanks to their green belts (hedgerows or tree lines), have an ecological value as green corridors: a synergy of ecosystem services ("win-win", TEEB 2010). Some stretches of road pass through the wood, so haven't a panoramic value, while others present a different mix of functions. When the hedgerows obtrude the view, the trade-off is a "some win/others lose" type.

This example shows that the trade-offs between landscape functions cannot be mechanically determined, but must be verified on a case-by-case basis. Some choices might be necessary, in the case of interventions, such as greening interventions, or, in contrast, tree-cuttings.

3.4.3 A Peri-urban Landscape: The Metropolitan Area of Turin

The metropolitan area of Turin, one of the largest in Northern Italy, presents a peri-urban landscape, still rich in open spaces with natural and cultural resources: natural parks, rivers, historical residences and gardens (including a UNESCO serial WHL Site), and minor traces such as historic farmhouses and channels. Despite their land capability, the remaining agricultural areas are poor and, not infrequently, abandoned. However, they are strategic for "framing" the above-mentioned resources, insofar as they constitute the landscape context of the built heritage, and essential ecological corridors.

For this reason, in 1999 the Piedmont Region implemented the "Corona Verde" strategic plan for protecting and enhancing the open spaces of the metropolitan area of Turin, from a landscape perspective (Cassatella 2013b). Thus, the Plan fully recognizes the multiple cultural functions that agriculture plays in a dense urban area. The scenic assessment supported this landscape project by mapping the rural areas which play a role in maintaining the legibility of historic and symbolic landmarks, and the openness towards important landscape frames in the background, such as the Alps (Fig. 3.2). The zone of visual influence of urban areas and infrastructure was mapped by GIS, and interpreted as indicator of perceptual disturbance on open areas.[7] This map also emphasizes their fragmentation.

With the help of European Regional Development Funds (ERDF), Corona Verde co-financed the creation of cycling routes and greenways, redevelopment and facilities for outdoor activities, renaturation and mitigation of infrastructural barriers, landscaping of the surroundings of heritage sites, awareness raising and promotional activities.

It should be noted that, until now, the Piedmont Region has made use of ERDF and not CAP incentives. Nowadays, it is clear that the challenge undertaken by Corona Verde needs a more integrated system of planning, implementation measures and financial instruments involving territorial, landscape and rural policies as well. We elaborate on this issue in broader terms in Sect. 3.5.3.

---

[7] This kind of interpretation is open to criticism insofar as urban areas include important landscape features (see Sect. 3.2).

Fig. 3.2 (*Above*) A bird's-eye view of the Turin metropolitan area. Natural and rural resources still coexist in the interface with the urban areas. (Photo: Archivio Direzione Regionale per i Beni Culturali e Paesaggistici del Piemonte). (*Below*) Main objectives of Corona Verde Strategic Plan (*Source* Politecnico di Torino 2007)

3.5 Supporting Landscape and Rural Policies Through Landscape Visual Assessment

3.5.1 Detailing Categories of Cultural Ecosystem Services Related to Landscape Amenity

Scenic quality supports several cultural services, related to non-use landscape values (identity, memory, spirituality, aesthetic enjoyment), and to recreational uses (outdoor activities, didactics) (Table 3.1). Moreover, thanks to landscape scenery, other environmental values can be "perceived" by the public, thus raising its awareness of them. Through the case studies, we observed several landscape functions of rural areas, namely: memory and aesthetics in an exceptional place, cultural values related to tradition and identity in an "ordinary" area, and recreation and ecological connectivity in a peri-urban context, where open spaces are a resource characterized by rarity. Each of these functions can be analysed through specific techniques, which can involve several disciplines (Daniel et al. 2012), in order to better understand the phenomena, the elements and the dynamics which have to be managed. The comprehension of their differences and specificities is essential for designing strategies and focusing on the appropriate techniques of interventions. For example, in the first case study (Albugnano) prescriptive regulations have been designed to protect well-defined scenes, while in the last one (Turin) an extensive knowledge system is provided to stimulate a variety of local actions within a strategic framework.

Scenic landscape assessment possesses a set of methods which can be useful for dealing with cultural services, identifying landscape values (including with reference to public opinion, by appropriate techniques not discussed in this chapter but see Grittani et al., *infra* Chap. 5 and Bottero, *infra*, Chap. 6), mapping the material and immaterial features which support perception, and defining quality objectives in order to contribute to policies. Above all, scenic assessment can highlight landscape features which play different functions, and potential interferences between environmental trade-offs. For example, in the case of San Martino Alfieri, the same element, a route, is, alternatively, a panoramic route or an ecological corridor, but cannot have both functions at the same time, because hedgerows and tree lines constitute, from a scenic point of view, a visual obstruction. A policy choice is needed. Only a sufficiently highly developed framework of cultural services, which recognizes specific landscape scenic values, can produce aware decision making about positive and negative trade-offs.

3.5.2 Developing Scenic Landscape Indicators for Environmental Assessment Frameworks

While landscape amenity is an ecosystem service and must not be confused with other kinds of cultural services, specific indicators are needed in the framework of landscape assessment and in the environmental assessment procedures where landscape is considered a component (such as EIA and SEA). As discussed in Sect. 3.2, the reduction of landscape assessment to only those aspects that can be measured by land-use or ecological indicators is as common as it is reductive and wrong from a theoretical perspective. Above all, when the assessment procedure regards policy effects, and not only landscape quality, such a reductionist approach risks underestimating the negative trade-offs between various functions, mostly in a multifunctional perspective.

The literature on landscape indicators of scenic quality (and even social perception) is rich in proposals and experimental case studies (Cassatella 2011). In this chapter, we have briefly illustrated visual sensitivity; in our view, VS can be applied to different types of landscape, rural, as well as urban and natural ones; it can be developed and mapped both at the local and at the regional scale; it is sensitive to modifications in land uses, and, as a consequence, can usefully inform decision-making and support landscape planning for the evaluation of alternative scenarios.

Moreover, VS can constitute a component variable for calculating other indexes related to scenic quality (illustrated in Sect. 3.3).

Finally, introducing scenic landscape indicators into current frameworks for evaluating rural, environmental and spatial policies can contribute to a stronger focus on landscape quality objectives, in line with the European Landscape Convention.

3.5.3 Integrating Rural and Spatial Policies to Protect and Enhance Landscape Scenic Beauty

In a given landscape scene, all of the elements are perceived in a holistic way, thus the scenic quality or rural landscape depends as much on fields and vegetation as on built elements, settlements and infrastructure. As a consequence, conservation and enhancement of rural landscape amenity can only be achieved by integrating rural and spatial policies and related tools, at any scale of intervention. In fact, a goal-oriented (or "normative", OECD 2001, p. 11) concept of multifunctionality implies multisectoriality.

Both landscape planning and spatial planning influence rural areas, in their open and built spaces. It is quite rare for ordinary planning to deal with scenery. However, a varied range of tools and experience offer interesting solutions for a wide range of situations and purposes (Table 3.3).[8] The first step is to establish landscape quality goals which explicitly recognize functions such as contemplation of aesthetics, recreation, conservation of memory, and so on. The second step is to compare these goals with current forms of land management, with trends, and with the scenario which is proposed by plans (with reference to all those plans with potential effects on landscape), in order to identify potential alliances and conflicts. The next step is to design strategies, plan measures and actions. The final step involves evaluating their implementation and effects. Scenic landscape assessment and landscape indicators are essential tools.

In the case studies, different planning strategies have been observed, mainly focused on the built environment. In fact, in the Italian context, spatial and landscape planning have little scope for regulating the rural open space, which is only subject to laws concerning agriculture and forestry as economic activities. Conversely, Rural Development Programmes (RDP) offer the chance to address a number of important drivers which shape the landscape, particularly via agri-environmental measures. In fact, the management of woods and forest, crops, hedgerows, even fallow land or set-aside areas, buildings and roads connected to rural practices may influence the landscape scene, its openness, sense of history and identity, accessibility and usability for recreational uses. As a consequence, the relationships between spatial planning and RDP must be strengthened (see also Rega, *infra*, Chap. 2).

In Italy, the Regions are responsible for implementing the RDP, the Regional Territorial Plan (RTP) and RLP. However, they are designed and assessed by separate departments of the regional administration. This causes a lack of coordination with regard to knowledge, goals, strategies, evaluative frameworks, and, ultimately, can lead to potential conflicts. The fact that all these instruments are subject to a common evaluation procedure, Strategic Environmental Assessment, may offer a window of opportunity to foster inter-departmental cooperation, harmonization of environmental objectives and knowledge sharing (see Rega, *infra*, Chap. 2).

At the regional level, the RDP, RTP and RLP could share:

- part of their knowledge apparatus (data and interpretations) on land uses, environment, ecology, and, of course, landscape;
- part of their strategic framework related to ecosystems and landscape, for example: the protection of traditional rural landscapes, or of landscape characterized by outstanding scenic beauty, can be a shared goal, based on a joint identification of target areas, which each plan or program will address by its own instrument, such as planning regulation or financial measures;
- part of their evaluation framework, mainly concerning landscape indicators. A shared database would contribute to a more efficient and economically feasible monitoring, and, mostly, to the identification of the multiplicity of values. For example, scenic landscape indicators could reveal positive and negative effects of greening measures on the scenery.

---

[8] See also Cassatella (2014) in press.

Table 3.3 Assessment techniques and planning tools for dealing with landscape services related to scenic values

Scenic values	Landscape quality objective(s)	Assessment techniques	Planning tools
Identity, due to the presence of landmarks	Protection of intervisibility between landmarks and vantage points; protection of prominence effect and symbolic hierarchies in the context	Identification and mapping of: zones of visual influence of landmarks; viewcones from vantage points; skylines; background	Protection acts; Building control (envelope, height, lightness), vegetation-cutting rules; incentives for open spaces near landmarks
Scenic amenity, due to visual openness	Protection of visual openness, by maintaining wide, deep views	Viewshed analysis and mapping; visual sensitivity mapping	Building control and limitations; tree-cutting rules for foreground; visual impact assessment for proposed developments in sensitive areas (avoiding obstruction, intrusion)
Distinctiveness of landscape characteristics, due to textures, colors, building types, natural elements (generally traditional character)	Conservation of landscape characteristics	Landscape Character Assessment; Landscape Atlases	Design codes; standards in local statutory plans; regulations on green spaces; voluntary and participative tools (Village Design Statements)
Tranquillity	Protection of tranquil areas	Mapping tranquil areas	Light/noise pollution measures
Integrity	Elimination or mitigation of detrimental features	Mapping detrimental features; their zone of visual influence and visual relationship with vantage points, landmarks and landscape assets	Planning requirements (elimination, remodeling, mitigation of detrimental elements); conditionality for new developments
Usability for recreational purpose (contemplation and outdoor activities)	Enhancement of systems for accessibility and equipment; conservation of syntactic sequences and symbolic relations among landscape assets and their surroundings	Mapping landscape resources for recreation; viewing places and routes (including minor and pedestrian ones; maritime, etc.); sequences of symbolic or historic value	Project strategies for improving accessibility and facilities; if appropriate, conservation measures for visual relationships and/or material characters of sites

At the local level, integration between rural programs and landscape and spatial plans could lead to integrated programs for designing multifunctional systems of green and open spaces, including agricultural parks, ecological networks and greenways. The plans may design the spatial scheme, while the rural programs may offer management criteria and financial instruments.

In Italy, an example of good practice is represented by the connection between the RDP of the Apulia Region, which foresees some landscape strategies of the RLP are supported by the funds of the CAP (Regione Puglia and Ministero per i Beni e le Attività Culturali 2013). In other contexts, such as the Piedmont Region, the RDP's agro-environmental measures barely address spatial targets, and this fact affect their effectiveness (Rega and Spaziante 2013). Just as environmental measures should target the most suitable areas, so should measures designed to improve the landscape quality and experience. In the latter case, "suitability" can be determined thanks to landscape assessment.

It should be noticed that the existence of a landscape plan is an important factor in designing multifunctional measures and in interpreting indicators, because it provides a landscape quality goals and spatial targets. In our view, multifunctionality is a goal-oriented concept, and the synthesis between amenity, productivity, and ecological services can only be achieved through a project-based approach.

In conclusion, whereas conserving or enhancing the character of rural landscape which support cultural services (because of its exceptional beauty, traditional character and identity, or recreational use or potential) is a policy goal, integrating rural, landscape and spatial policies is necessary, on the basis of horizontal collaboration and synergy between sectors of public administration, so as to ensure the sharing of knowledge, goals, projects, targets, evaluations and financial instruments.

References

Agnoletti M (ed) (2012) The italian historical rural landscape. Cultural values for the environment and rural development. Springer, New York

Antrop M (2007) Reflecting upon 25 years of landscape ecology. Landscape Ecol 22:1441–1443

Brinkhuijsen M (2007) Legible landscapes: the use of narratives in landscape design for leisure and tourism in Dutch cultural landscapes. In: Selwyn T, Scott J (eds) Proceedings of the ASA conference 2007, thinking through tourism, London Metropolitan University, 10–13 April 2007

Cassatella C et al (2009) Consumo di suolo, consumo di paesaggio? prospettive di ricerca sulla misura delle ricadute dei programmi di Sviluppo Rurale, VA. VALUTAZIONE AMBIENTALE, 16:12–17

Cassatella C (2014) Landscape scenic values: protection and management from a spatial-planning perspective. In: Gambino R, Peano A (eds) Nature policies and landscape policies: towards an alliance. Springer, Dordrecht (in press)

Cassatella C (2013a) Guidelines for the analysis, protection and enhancement of landscape scenic characters, Politecnico di Torino-DIST, 2012–2013. Research report commissioned by Italian Ministry for Cultural heritage and Activities, Piedmont's Regional Direction

Cassatella C (2013b) The 'Corona Verde' Strategic Plan: an integrated vision for protecting and enhancing the natural and cultural heritage. Urban Res Pract 6(2):219–228. doi:10.1080/17535069.2013.810933

Cassatella C (2011) Assessing visual and social perceptions of landscape. In: Cassatella C, Peano A, Landscape indicators. Assessing and monitoring landscape quality. Springer, Dordrecht. doi:10.1007/978-94-007-0366-7_6

Cassatella C, Voghera A (2011) Indicators used for landscape. In: Cassatella C, Peano A (eds) Landscape indicators. Assessing and monitoring landscape quality. Springer, Dordrecht, pp 31–46

Cassatella C, Seardo BM (2012) I valori scenico-percettivi. In: Larcher F (ed) Prendere decisioni sul paesaggio. FrancoAngeli, Milano, pp 66–79

Constituent Assemby of Italy (1947) Constitution of the Italian Republic, approved on 22nd December 1947

CoE (Council of Europe) (2000) European Landscape Convention, Florence, CETS No.: 176

Churchward C et al (2013) Evaluation of methodologies for visual impact assessments. In: US TRB's National Cooperative Highway Research Program (NCHRP) Report 741, Washington. http://www.trb.org/Main/Blurbs/168536.aspx. Accessed May 2013

Daniel TC (2001) Whiter Scenic Beauty? Visual landscape quality assessment in the 21st century. Landscape Urban Plan 54:267–281

Daniel T et al (2012) Contributions of cultural services to the ecosystem services agenda. PNAS 109:8812–8819

DEFRA—Department of the Environment, Food and Rural Affairs (2000) Rural white paper. Our countryside: the future. Defra Publications, London

van der Heide C, Heijman W (2013) The economic value of landscape. Routledge, New York

Italian Republic (2012) Decree n. 17070 of 19 November 2012 by the Ministry of Agriculture Food and Forestry Policies for the institution of the "National Observatory of Rural Landscape, Agricultural Practices and Traditional Knowledge"

Italian Republic (2004) Codice dei Beni Culturali e del Paesaggio (Italian Code for Cultural Heritage and Landscape), Decree No 42 approved on 22nd January 2004 (and subsequent modifications and integrations)

Landscape Institute and Institute of environmental Management and Assessment (2013) Guidelines for landscape and visual impact assessment. Routledge, New York

Lynch K (1960) The image of the city. MIT Press, Cambridge

Millennium Ecosystem Assessment (2005) Ecosystems and human well-being: synthesis. Island Press, Washington

NE (Natural England) (2009) Experiencing Landscapes: capturing the cultural services and experiential qualities of landscape. Natural England Commissioned Report NECR024, Sheffield

Ode A, Tveit MS, Fry G (2008) Capturing landscape visual quality character using indicators: touching base with landscape aesthetic theory. Landscape Res 33(1):89–117. doi:10.1080/01426390701773854

OECD (Organisation for Economic Co-Operation and Development), (2001) Multifunctionality: towards an analytical framework. Agriculture and food, Paris

Politecnico di Torino—Dipartimento Interateneo Territorio (DITER)—Politecnico e Università di Torino (2007) Progetto Corona Verde: pianificazione strategica e governance report. http://www.ocs.polito.it/coronaverde07.htm

Rega C, Spaziante A (2013) Linking ecosystem services to agri-environmental schemes through SEA: a case study from Northern Italy. Environ Impact Asses 40:47–53

Regione Puglia and Ministero per i Beni e le Attività Culturali, Direzione Regionale per I Beni Culturali e Paesaggistici della Puglia (2013) Piano Paesaggistico Territoriale—Regione Puglia, approved by the Regional Committee on August 2nd

Rentig H et al (2009) Exploring multifunctional agriculture. A review of conceptual approaches and prospects for an integrative transitional framework. J Environ Manage 90:S112–S123

Seardo B (ed) (2012), Planning multifunctional landscapes: from large-scale to project. Urbanistica n 148:75–82

TEEB (2010) The Economycs of Ecosystems and Biodiversity: Mainstreaming the economics of nature: a synthesis of the approach, conclusions and recommendations of TEEB

Swanwick C (2002) Landscape character assessment. Guidance for England and Scotland, prepared on behalf of The Countryside Agency and Natural England

UNESCO (1998) Convention concerning the protection of the World Cultural and Natural Heritage, World Heritage Committee-Twenty-first session, Naples-Italy 1–6 December 1997, Paris, France

USDA-FS (Unites States Department of Agriculture—Forest Service) (1995) Landscape aesthetics. In: A handbook for scenery management, agriculture Handbook No. 701

USDI-NPS (United States Department of the Interior—National Park Service), Yosemite National Park (2010) Scenic vista management plan for Yosemite National Park, environmental assessment. Yosemite, California

VROM (Dutch Ministry of Housing, Spatial Planning and the Environment) (1999) The Belvedere Memorandum. A policy document examining the relationship between cultural history and spatial planning. VROM, Den Haag

VROM (Dutch Ministry of Housing, Spatial Planning and the Environment), LNV (Dutch Ministry of Agriculture, Nature and Food Quality) (2008) Agenda landschap—Landschappelijk verantwoord ondernemen voor iedereen. VROM, Den Haag

Chapter 4
A Multi-scale Approach to Support Integrated Landscape Management in Rural Mountainside Areas (RMAs) of Alps

Chiara Bragagnolo, Chiara Rizzi and Stefania Staniscia

Abstract Agriculture has historically played a fundamental role in shaping and conserving mountainside landscapes of Alps. Since the emergence of the sustainability principles and following the reform of the EU agricultural policies, a multifunctional role of the primary sector has been increasingly recognised, especially in those contexts where farming practices have greatly contributed to preserve social and environmental capitals. Referring to mountain landscapes, the effects on the environment caused by the abandonment of agro-pastoral practices due to both physical constraints and economic factors have extended beyond the local scale, changing landscape characters and cultural traditions of Alpine regions. Thus, a greater integration among sectoral policies has been often recommended in order to enhance alpine rural areas, traditions and products. However, changes of rural landscapes and the abandonment of agricultural practices has not occurred likewise across the Alps, as mountainside areas present different values (economic, ecologic, etc.) and they are exposed to different risks (human pressures, hydrogeology, etc.). In addition, institutional and non-institutional initiatives for protecting mountainside areas have been implemented differently across the Alps according to particular landscape management and governance systems of countries and regions. With this contribution, we aim to present the results of a research project funded by the Autonomous Province of Trento (Italy), aiming to support landscape management in Rural Mountainside Areas (RMAs) of Alps. A multi-scale approach is developed and applied to the Province of Trento, a mountain region located in the southern side of the Alps (Northern Italy) which

C. Bragagnolo (✉)
Institute of Biological and Health Sciences, Federal University of Alagoas,
Av. Lourival Melo Mota, s/n, Tabuleiro do Martins, Maceió, AL 57072-900, Brazil
e-mail: chiara.bragagnolo@ing.unitn.it

C. Rizzi · S. Staniscia
Department of Civil, Environmental and Mechanical Engineering (DICAM),
University of Trento, Via Mesiano 77 38123 Trento, Italy
e-mail: chirizzi@gmail.com

S. Staniscia
e-mail: stefaniastaniscia@yahoo.it

C. Rega (ed.), *Landscape Planning and Rural Development*,
SpringerBriefs in Geography, DOI: 10.1007/978-3-319-05759-0_4,

benefits from a special Autonomous Statute. The method was developed following an interaction process based on internal and external meetings with experts organised during the project. It integrates quantitative (maps and data) and qualitative (field works, expert opinions, etc.) information and analyses. Firstly, RMAs are defined and identified within the study region. Then, five priority contexts are selected based on the assessment of values and risks. They correspond to those RMAs reflecting a significant combination of risks and values. Finally, integrated measures are proposed in order to support landscape management in priority contexts. The results show the importance of considering broader scale values and risks in order to develop and implement operative measures for preserving rural landscapes in mountainside areas. Based on lessons learned, a greater cooperation among authorities, sectoral offices, researches, local experts and public is strongly recommended to integrate landscape issues into spatial and sectoral decisions.

Keywords Rural Mountainside Areas (RMAs) • Landscape management • Multi-scale approach

4.1 Introduction

The European Landscape Convention (hereafter ELC) defines *landscape management* as an action ensuring the regular upkeep of a landscape, guiding and harmonising changes (CoE 2000). Agriculture has historically contributed to landscape management, by shaping landscape forms and characters for subsistence and productive purposes (Sereni 1961; Antrop 2005). Mountain farming is one of the most representative results of the action and interaction between people and environment, reflecting a long-term relationship that progressively changed an adverse landscape into a suitable source of food and income for mountain populations (Franceschetti and Argenta 2002; Pôças et al. 2011; EU Mountain Farming Protocol 2006).

Since the emergence of the sustainability concept, a multifunctional role of primary sector has been internationally recognised and increasingly imposed as new paradigm for European agriculture and rural development (Huylenbroeck and van Duran 2004). Moreover, the multifunctional role of agriculture has been particularly emphasised in respect of mountain areas where extensive farming practices have greatly contributed to: biodiversity and habitats conservation, natural hazard prevention, water management, food security, among others (Ferrario 2012).

In Europe, mountain regions present a rich variability of rural landscapes (Plieninger et al. 2006), conserving high biodiversity and cultural values associated with traditional and ecological knowledge built upon different physical adverse environments (Rescia et al. 2008). Terraced agricultural landscapes are the testimonial of this richness, showing how agro-pastoral intervention has been able to adapt to different altitudes, slopes, aspects and climates (Fontanari and Patassini 2008). However, during the last decades, European rural landscapes have been experiencing homogenisation and fragmentation processes (Jongman 2002) and mountain rural landscapes have been largely abandoned, reflecting a depopulation trend associated with remote and less competitive areas (MacDonald et al. 2000) and changing perceptions and values of peoples (Antrop 2005). In this context, the effects on the environment caused by the abandonment of agro-pastoral practices due to both physical constraints and economic factors have extended beyond the local scale, changing landscape characters and cultural traditions of Alpine regions (Haddaway et al. 2013). Thus, during the last decades, the abandonment of mountain agricultural landscapes in the alpine region has raised the attention of institutions, researchers and population due to the increase of natural hazards and the loss of traditional cultivation forms, biodiversity and cultural heritage (Lodatti and Patassini 2008).

However, changes of rural landscapes and the abandonment of agricultural practices has not occurred likewise across the Alps, as mountain regions present high variability (cultural, socioeconomic and environmental) and have passed through different development processes leading to heterogeneous spatial and landscape patterns (Diamantini 1996). In addition, institutional and non-institutional initiatives for protecting mountain areas have been implemented differently across the Alps according to particular landscape management and governance systems of countries and regions (Morschel et al. 2004).

Although there is not a general recipe for preserving mountain regions and their rural landscapes, a greater integration among territorial and sectoral policies (i.e. landscape preservation, rural development, spatial planning, water management, etc.) has been largely advocated to protect and recuperate rural areas, traditions and products (Claval 2005; Tassinari 2008). Integration is a key principle of the ELC, which suggests that landscape protection needs to be an integral part of territorial and sectoral policies with landscape implications (CoE 2000). However, the integration of policies is essential but not enough to achieve an effective landscape management since it further requires extensive co-ordination at all levels of decision-making (CoE 2006).

According to the ELC, innovative instruments and tools for landscape management should be developed and established based on the fact that landscapes (including mountain rural landscapes) are changing components. Therefore, their protection, planning and management could benefit from actions supporting the assessment of their state and setting up monitoring systems to follow and manage those changes (CoE 2000; Antrop 2005). Moreover, inventorying and monitoring small-scale landscape changes and elements composing landscape characters may be crucial to preserve broader scale values and prevent broader scale risks such as the abandonment of mountain rural landscapes.

In this chapter we present a multi-scale approach to support integrated landscape management in Rural Mountainside Areas (RMAs) of Alps. The approach was developed within a research project funded by the Autonomous Province of Trento (Italy). The Province of Trento is a mountain region located in the southern side of the Alps (Northern Italy) which benefits from a special Autonomous Statute. It is divided into 217 municipalities and 16 Communities (a new administrative level introduced by a recent reform). The region presents heterogeneous climates and environment conditions (from glaciers to plains with sub-

Mediterranean climate), hosting a high variability of natural and rural landscapes, cultural identities and land use patterns.

In 2010, the Office of Regional and Town Planning and Landscape Protection of the Province of Trento (*Servizio Urbanistica e Tutela del Paesaggio*) opened a public call to finance a series of thematic research projects on landscape. The call aimed to generate knowledge and methods to support the future development of regional landscape planning. A research proposal aiming at developing a methodology to analyse and manage RMAs was presented and, after an assessment process, it was commissioned to our multidisciplinary project team. The research project started in July 2011 lasting one year.

The methodology integrated quantitative (maps and data) and qualitative (field works, expert opinions, etc.) analyses, and it was developed following an interaction process based on external and internal meetings with experts organised during the research period.

Firstly, we defined and mapped RMAs at large scale, creating a *census database* based on top-down (maps) and bottom-up (field works) analyses (214 areas were detected and described). Secondly, we selected priority contexts (five RMAs) based on the assessment of values and risks. Finally, we adopted a matrix-based approach to define integrated measures to support landscape management in priority contexts.

The Chapter is structured as follows: Sect. 4.2 describes the conceptual and methodological background of this research. Section 4.3 focuses on the case study region, providing a brief portray of: geopolitical context, historical evolution of primary sector and rural landscape, and relevant policy and regulatory instruments for landscape management. Section 4.4 briefly tells the story of this research project. Section 4.5 describes the multi-scale approaches, including methods used and main results obtained. Finally, Sect. 4.6 discusses key methodological issues and practical aspects, providing concluding remarks and highlighting the main lessons learned.

4.2 Contextual and Methodological Background

Principles and aims of both the European Landscape Convention (ELC) and the Alpine Convention (AC) were the main conceptual basis of this project and the ALPTER EU-Interreg project has helped in developing the methodological approach described in Sect. 4.4.

The ELC—signed in Florence on 20 October 2000—introduces a Europe-based definition of landscape (*area whose character is the result of the action and interaction of natural and/or human factors*, cf. ELC—art 1), basing on the quality of protection, management and planning of the entire territory, not just exceptional landscapes. This "democratisation of landscape" (Priore 2000) represented a great innovation since it extends the definition of landscape to the entire European territory, including everyday or degraded landscapes and acknowledging a social preference to them. Accordingly, rural areas are formally included as part of the landscape and farming activities are recognised as interactive actions contributing to create cultural identity and shape landscape character (cf. ELC—art 2). Among others, the Convention calls for the contracting Parties to integrate landscape into its regional and town planning policies and in its cultural, environmental, agricultural, social and economic policies, as well as in any other policies with possible direct or indirect impact on landscape (cf. ELC—art 5, paragraph d). This integration principle must guide landscape actions, developing and establishing appropriates tools and increasing the cooperation among policies, researches and policy makers (CoE 2008).

The Alpine Convention is an international treaty for the sustainable development of the Alps signed in Salzburg on 7 November 1991 by 8 Alpine countries—Austria, France, Germany, Italy, Liechtenstein, Monaco, Slovenia and Switzerland—and the European Union (its ratification has terminated on 27 March 2000). The objective of the treaty is to protect the natural environment of the Alps while promoting its development. Under the Convention, contracting Member States should adopt specific measures in 12 thematic issues (population and culture, spatial planning, air pollution, soil conservation, water management, conservation of nature and the countryside, mountain farming, mountain forests, tourism, transport, energy, and waste management). Of these issues, eight protocols were elaborated and annexed to the Framework Convention. Of particular concern for this research is the Mountain Farming Protocol (concluded in Chambéry on 1994 and adopted on 2002), implementing the Alpine Convention in the field of mountain farming. On the one hand, it formally recognises the multifunctional role of mountain agriculture, stating *the importance which farming has always had in the Alpine region and the indispensable contribution which this branch of the economy makes, and will continue to make, particularly in mountain regions* and stressing on the fact that *farming methods and intensity exert a decisive influence on nature and landscapes and that extensively farmed countryside must fulfil an essential function as a habitat for Alpine flora and fauna*. On the other hand, it recognises *that the geomorphology and climate of mountain regions create more difficult living and production conditions for farming activity*, suggesting the implementation of a cross-border framework and jointed measures to ensure that *the future of farmers and their farms in mountain regions is not called into question by the application of exclusively economic parameters* (EU Mountain Farming Protocol 2006).

The project ALPTER was co-financed in the framework of EU programme *Interreg Alpine Space* with the aim to counteract the abandonment of terraced agricultural areas in the alpine region. It started in 2005, portraying the state-of-the-art of scientific knowledge about terraced landscapes and defining procedures with regard to significant topics such as mapping, assessment of geological hazard, enhancement of agricultural production or promotion of tourism in terraced zones. A series of best-practices based on integrated strategies and actions (from tourism promotion to agricultural production) were further included, demonstrating the importance of alpine terraced agricultural areas as well as the compelling need to define strategies and solutions based on integrated management to preserve them. The ALPTER project has further suggested that recognising the values of these areas (such as the mitigation of hydrogeological risks, the protection of biodiversity and the sustainability of the agricultural production) is essential but not sufficient to support the recovery and restoration of these landscapes (Fontanari and Patassini 2008). Then, two recommendations are given in order to improve landscape planning of terraced agricultural areas: to increase knowledge and data on these unique landscapes and to identify key actors and develop operative interventions for enhancing and restoring terraced landscapes.

4.3 The Province of Trento

4.3.1 Geo-political Context

The Province of Trento (Trentino) is a mountain region located in the southern side of the Alps (Northern Italy) (Fig. 4.1). It covers a surface of 6,207 km^2 and 70 % of the total area is up to 1,000 m above sea level.

The region is made up by several alpine and subalpine valleys presenting heterogeneous climate and environment conditions (from glaciers to plains with sub-Mediterranean climate). This leads to a great variability of natural and rural landscapes, cultural identities and land use patterns. The region is mainly covered by forests (53 %) and natural areas (31 %), and agricultural value added represents 2.8 % of the regional economy (PAT 2012a). Primary sector is mainly represented by extensive managed farmlands which have encountered the location where natural and climate conditions were more favourable (e.g. elevation, slope, aspect, etc.), playing a key role in shaping the landscape of the hillsides of valleys.

The Province benefits from a special Autonomous Statute, giving it the legislative and administrative control over a wide range of sectors (e.g. regional economy, environmental subjects, implementation of international agreements and EU rights and policies at local scale, cultural issues, etc.). According with a recent reform, 16 Communities have been identified and officially designed as intermediate institutions between the Province and the 217 local municipalities, giving them the administrative control over: spatial planning, social welfare, housing policies, and coordination of local issues and municipalities (Fig. 4.1).

4.3.2 The Historical Evolution of Primary Sector and Rural Landscape in Trentino

During the second half of the nineteenth century, the economy of Trentino was largely based on small-scale agriculture and livelihood production. Since the first industrial revolution, the economy has passed through several crises, driven by the combination of market-based and socio-environmental factors (i.e. unspecialised crops, poor competition with large-scale production, huge fragmentation of farmer

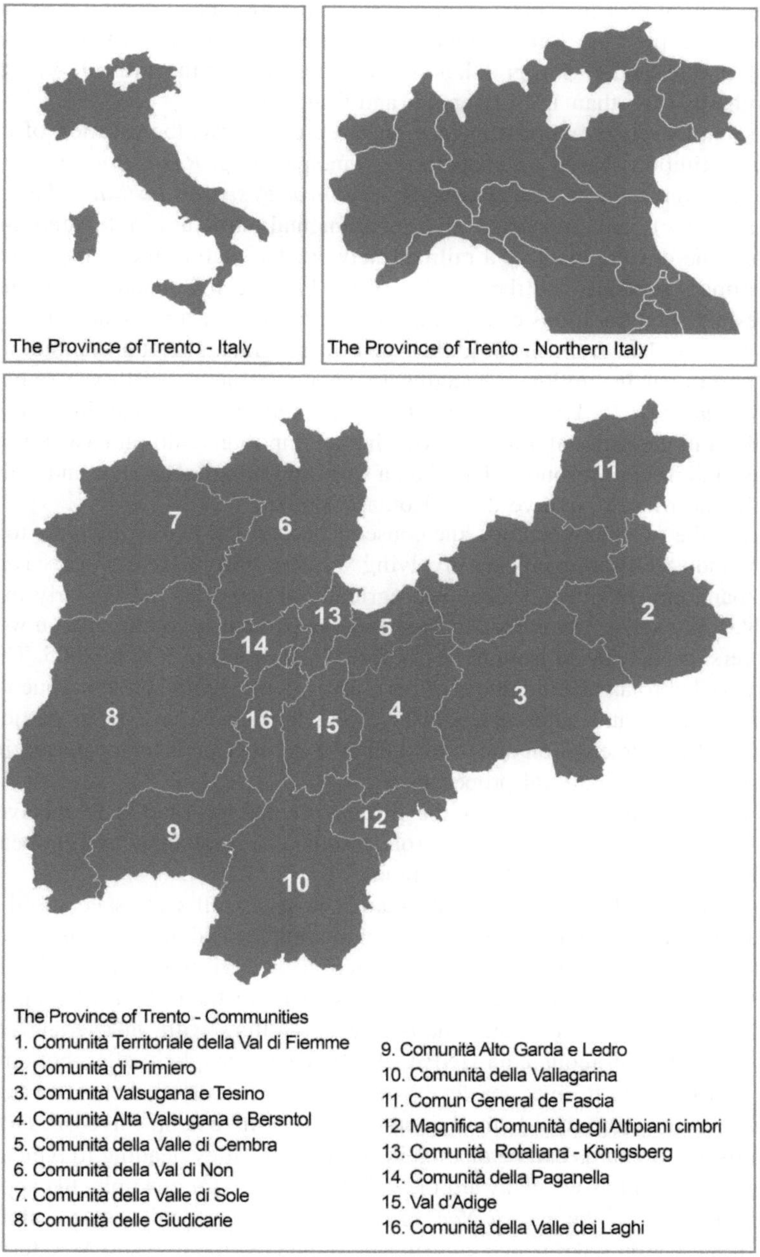

Fig. 4.1 The Province of Trento and its communities

tenures, natural disasters, plant diseases, etc.). In 1900, the agricultural productive area was occupied by: forest (about 50 %), mountain and non-mountain pasture (25 %), arable land (7 %), grassland (7 %), specialized vineyards (1 %), gardens and orchards (less than 1 %) (Bonoldi and Cau 2011).

The main strategy adopted to face on these crises was the creation of collaborative institutions based on cooperative principles (e.g. *Rural Bank of Trentino*, *Provincial Agricultural Consortium*, *Agricultural Research Institute of San Michele all'Adige*) and supported by governmental structural interventions (for example, the development of a railroad network facilitating the transportation of local primary products and the creation of the Provincial Council of Agriculture). Long-established traditions of community forestry and pasture management in the north of Italy date from the Middle Ages. In several mountain communities of Alps, traditional institutions managing mountain resources still exist (e.g. *Magnifica Comunità della Val di Fiemme*, *Regole D'Ampezzo*, etc.), carving out special exceptions in the national law and providing an important cultural basis for the use of forest and pasture resources based on a spirit of mutual assistance and solidarity (Spinale and Manez Archive 2002; Kothari 2006).

During the twentieth century, the consequences of the two world wars together with the industrialisation process involving Western Europe have progressively led to an abandonment of rural areas and agricultural activities, particularly in unfavourable contexts as mountains. In Trentino, the percentage of population working as farmers has decreased from 50 % (before world wars) to 5 % in 2005. This has inexorably led to an abandonment of rural areas, with significant consequences on the environment and landscape. Furthermore, farmlands have been particularly vulnerable to fragmentation due to the lack of legal mechanisms for preserving the indivisibility of agricultural properties.

However, during last decades the primary sector has started to receive new attention, due to the multifunctional role assigned to rural areas by European and Provincial Rural policies and programmes.

Nowadays, the Province of Trento mainly hosts: small-scale specialised farms producing high quality products (mainly wine, apples and small fruits), and high altitude and larger farms with a main function of environmental and landscape protection (mainly pasture). The first represent more than 70 % of total farms, characterised by extensive and small-scale agriculture with an average utilised agricultural area of about 2 ha (INEA 2012). They occupied valley bottoms and hills (until 600/700 m over sea level) and they are mostly dedicated to the production of wine, apples and small fruits, and, to some extent, olive oil. Different vineyards and orchards can be distinguished in each valley, leading to typical high quality local products which award the valley identity—for example, the apple has become the icon of the *Val di Non*, and olive groves are only present along the edges of the Garda Lake which contributes to the creation of a sub-Mediterranean micro-climate favouring the production of a typical olive oil. These cultivations are mostly stretched over terraced hill slopes shaped by hydraulic interventions and agricultural practices giving a peculiar landscape character. The presence of small rural buildings and assets (e.g. drywalls, water channels, etc.) as well as singular cultivation forms (e.g. *pergola* or *arbour*s—a cultivation forming a shaded walkway of vertical pillars that usually support cross-beams upon which woody vines are trained) are usual. Moreover, the primary cultivation (vineyards or orchards) is often combined with livelihood productions (livestock, vegetables, wood, etc.) and semi-natural elements (vegetation remnants, hedgerows, etc.) conferring quality to the agricultural landscape.

Moreover, the importance given to the multifunctional role of mountain agriculture has led to the increase of agritourism activities and enterprises. From 2003 to 2009, farms offering tourism services increased by 74 % (PAT 2012b), offering accommodations, farm stays, bed and breakfasts and restaurant services, camping areas and, to some extent, educational programmes (mainly offering children opportunities to learn how a farm functions).

4.3.3 Policy and Regulatory Instruments for Landscape Management

Many instruments can be implemented to preserve rural landscapes. This paragraph focuses on those implemented at regional scale mostly influencing rural landscape planning and management: regional spatial planning, the Rural Development Programme (2007–2013) and the landscape fund.

4.3.3.1 Regional Spatial Planning

The current spatial plan of the Province of Trento has been approved in 2008, shifting from a land-use regulation instrument to a strategic plan, integrating spatial and sectoral policies to orientate the future development of the regional and local contexts (i.e. Communities and municipalities). As a result, it includes landscape and rural strategies based on sustainable development principles and long-term objectives.

Landscape issues have been regularly considered by regional spatial plans of the Province. The first (approved in 1967) considered the landscape a public good, declaring the need to preserve it in order to increase, among others, tourism-based activities. The second (approved in 1987) considered landscape protection a crucial way to preserve environmental values and contain urban sprawl. The current spatial plan (PAT 2008) introduces the concepts and principles of the European Landscape Convention, considering landscape protection as a way to preserve landscape characters, environmental quality and cultural identity.

Referring to agriculture and rural development, the spatial plan acknowledged some threats, including the urban encroachment of rural areas and the degradation of rural landscapes associated to the abandonment of agricultural practices.

The most important planning instruments for landscape management are: the development of a provincial landscape map and the establishment of stricter restrictions to preserve landscape characters. The first identifies different landscape-related contexts and elements based on their geophysical and socio-cultural values (e.g. traditional buildings and settlements, rural areas, alpine landscapes, etc.). It aims to support regional and lower scale decisions concerning the protection of landscape values and the definition of landscape interventions and transformations. The second identifies high quality landscapes and unique elements (including high quality rural areas), where stricter regulations should be established in order to preserve them (for example more severe regulations on building and urban development, etc.).

4.3.3.2 The Rural Development Programme of the Province of Trento

According to the National Strategic Plan for agricultural and rural development (elaborated within the EU Rural Development policy 2007–2013), the Province of Trento is classified as "area with complex problems concerning development" (MIPAAF 2010).

The Rural Development Programme (hereafter RDP) (2007–2013) (PAT 2013) has started from this perspective in order to meet the objectives established by the EU rural policy (see Rega, *infra,* Chap. 1).

Two of the three main axes of the programme are considered relevant for this research:

- Axis II—Improvement of environment and rural areas, by implementing sustainable management measures;
- Axis III—Quality of life in rural areas and diversification of the rural economy, based on the implementation of integrated management strategies encouraging the diversification of rural economy and activities.

The axes establish financial incentives and mechanisms aiming at: compensating low productivity, paying for environmental services (e.g. biodiversity conservation, etc.), incentivising the production of organic products and eco-friendly agro-practices, promoting agritourism and education activities, restoring rural landscapes and buildings, among others.

According to the recent framework delineated within the new programming period for Rural Development (2014–2020), the abandonment of rural activities in marginal areas seems to have not stopped, leading to a disordered reforestation as well as a homogenisation of traditional composite landscapes. Thus, the new RDP plans to follow the objectives of the precedent period, with the aim to contain this non-stopped phenomenon.

4.3.3.3 The Landscape Fund: A Financial Resource for Managing Rural Landscape at Regional Scale

The landscape fund represents a core resource administered by the Office of Regional and Town Planning and Landscape Protection of the Province of Trento (*Servizio Urbanistica e Tutela del Paesaggio*) for financing projects and interventions on both site specific historical settlements and broader scale landscapes. It was introduced in 2008 by the Provincial Urban Act (LP n.1/2008 art.78). Its main objective is to enhance and restore historical settlements and broader scale landscapes by financing local authorities, public and private institutions, enterprises and agencies. The financial budget is established yearly within the economic budget approved by the Provincial Authority.

4.4 The Research Project

This project was financed by the landscape fund of the Province of Trento. In 2010, a public call was opened in order to finance eight thematic proposals embracing strategic issues for the future development of regional landscape planning. The call financed broader studies (e.g. analysis of the evolution of the landscape), methodological proposals (requalification of abandoned mines, building techniques for retrofitting rural artefacts, etc.) and site-specific projects (restoration of scenic landscape of the Dolomites passes, etc.).

A research proposal was presented within the call's theme *Methodological approaches for the transformation of rural mountainside areas through innovative production techniques*. The main objective of the proposal was to develop a methodology for supporting the analysis and the management of Rural Mountainside Areas (RMAs).

Following an assessment process lasted 6 months, the work was commissioned in honour of our project team, headed by a private architecture firm (Studio Ricci and Spaini). The project team was multidisciplinary, including landscape and urban planners, environmental designers, architects, local experts, agronomists, geologists and environmental scientists.[1] Several of them have also an academic position.

The assessment report stated that the research proposal recognised the multiple values of rural mountainside areas and related assets (in particular terraces) as well as the risks associated with important changes affecting those landscapes (e.g. mechanisation of agricultural practices and techniques, climate change, etc.). This

[1] Arch. F. Spaini (reaserch leader), Prof. Arch. M. Ricci (landscape and urban planner), Arch. PhD C. Rizzi (architect and expert on environmental design), Arch. PhD S. Staniscia (landscape architect), PhD C. Bragagnolo (environmental scientist and GIS expert), A. Gelmetti (agronomist and local expert), Arch. PhD M. Ferretti (architect), Arch. F. Pontalti (architect and local expert), C. Belloni (geologist), Arch. M. Bonvecchio (young professional trainee).

received particular interest due to both the lack of information on rural mountainsides areas at regional scale and the compelling need to find management instruments to preserve them based on their values.

The research project started on July 2011 and lasted 1 year. Six meetings with external experts were organised between July and November 2011, strengthening the cooperation between the project team, decision-makers (i.e. the sectoral Offices of the Province) and technical institutions (Committee of Agricultural Areas, Edmund Mach Foundation, etc.). Furthermore, internal meetings were organised monthly, favouring the exchange of information, knowledge and skills within the project team.

This chapter shows only part of the project results.[2]

4.5 The Multi-scale Approach

To support integrated management in rural mountainside areas we developed a multi-scale approach, including the following steps:

1. definition and localisation of RMAs;
2. selection of priority contexts;
3. definition of integrated measures.

The methodology integrated quantitative (maps and data) and qualitative (field works, expert opinions, etc.) analyses. Available spatial data applied to develop maps and GIS-based analysis were provided by different offices of the Province, supporting the overall project and facilitating the discussion among experts.

4.5.1 Definition and Localisation of RMAs

The first challenge of the project was to provide a definition of Rural Mountainside Areas (RMAs) since general definitions were lacking in literature. After the first two meetings, we defined these areas as:

> Rural landscapes located on hill or mountain slopes (generally terraced) shaped by hydraulic interventions and extensive agricultural practices conferring a high socio-cultural value.

Several common features characterising RMAs were further identified, including:

[2] The extensive results of this project can be found at: http://www.paesaggiotrentino.it/it/progetti-e-azioni/progetti-fondo-del-paesaggio/4.-indicazioni-metodologiche-per-le-trasformazioni-delle-aree-agricole-di-versante-a-nuove-tecniche-produttive-studio-ricciespaini_4202_ids

- the presence of terraced slopes and/or man-made rainwater systems (e.g. wells, tanks, etc.);
- primary land use cover (mainly vineyards and orchards—olive groves only in a southern valley);
- secondary land use cover (mainly semi-natural and natural areas);
- low soil fertility and depth;
- traditional cultivation forms with high landscape and identity value.

To map RMAs at large scale, we combined top-down and bottom-up analyses. The first adopts a spatially explicit approach, providing a map of the most probable distribution of RMAs at regional scale. It was based on the combination of spatial criteria describing the main features of RMAs pointed out (Box 1). The second used expert opinions and field research (field data collection, pictures, etc.) to identify a list of RMAs within the region (about 200 areas were identified during the first meeting and a list of 214 RMAs was agreed in the second meeting). Then, these areas were detected from aerial imagery and mapped.

Additional information collected during field works concerned: the presence of peculiar landscape elements (rural buildings and assets, drywalls, terraces, etc.), the management state (cultivated, in transformation, abandoned), land use cover properties, visibility from the area, and positive and negative elements contributing to landscape quality (e.g. presence of sprawled settlements, degraded water systems, etc.). This information was collected and used in all steps of the project.

Box 1: Mapping RMAs at Large Scale

The methodology developed to map the most probable distribution of RMAs at large scale included the following steps:

(i) selection of relevant criteria describing RMAs;
(ii) assignment of a probability to each criterion;
(iii) development of a probability map of RMAs at large scale based on the combination of criteria.
(iv) Selection of relevant criteria describing RMAs

The criteria were defined and selected based on expert opinions collected during project meetings, including:

1. Elevation (C1): it considers that RMAs are usually allocated between 75 and 1,000 meters above sea level;
2. Slope (C2): it reflects the fact that RMAs slope generally ranges between 5 and 50 %;
3. Aspect (C3): it considers that RMAs are preferentially located in south-facing slopes receiving more hours of sunlight;

4. Land use cover (C4): it selects a set of representative classes for RMAs, namely: heterogeneous crops; orchards, olive groves, vineyards, farmlands and uncultivated arable lands.

The four criteria were mapped from the spatial data provided by Province's Offices, including: Digital Terrain Model (DTM) derived from LIDAR (a remote sensing technology measuring distance with a laser used to make high resolution maps) and Land Use Map (scale 1:10,000).

(i) Assignment of a probability to each criterion

Two probability levels were assigned to each criterion (low and high) based on their suitability to describe RMAs location (thresholds were established based on expert opinions collected during project meetings). Therefore, a greater probability was assigned to:

1. the elevation ranges between 75 and 750 m above sea level (C1);
2. the slope ranges between 5 and 35 % (C2);
3. south-facing aspects (C3);
4. the following classes of land use covers: heterogeneous crops; orchards, olive groves, vineyards (C4).

(ii) Development of a probability map of RMAs at large scale

To develop a probability map the criteria were combined applying a simplified algorithm. The map shows the most probable spatial pattern of RMAs at large scale.

The results of both analyses showed how the majority of RMAs are concentrated in 10 Valleys, namely: Alta Valsugana e Bersntol, Valsugana e Tesino, Valle Alto Garda e Ledro, Valle delle Giudicarie, Valle Rotaliana-Königsberg, Valle dell'Adige, Vallagarina, Val di Cembra, Val di Non, Valle dei Laghi (Fig. 4.2).

A synthetic description of them was provided by calculating the criteria applied to develop the probability map for each valley. This allowed valleys to be compared in terms of average slope, average elevation, primary land use, etc. Moreover, by overlapping the polygons of RMAs detected from aerial imagery with those criteria, an inventory of RMAs (or census database) made up by 214 charts was created. Firstly, RMAs were classified according to their location, by assigning a letter code to each valley (e.g. AG for Alto Garda) and associating a progressive number to each RMA belonging to that valley (e.g. AG4). Thus, each RMA received an ID code. Then, key features were calculated through basic GIS-based functions (e.g. zonal statistic; area calculation, etc.) for each RMA. Each chart described each RMA in terms of relevant features (e.g. surface, average slope, maximum and minimum elevation, average solar aspect, land use cover, etc.).

Fig. 4.2 Rural mountainside areas in five selected valleys

4.5.2 Selection of Priority Contexts

The selection of priority contexts was based on the assessment of values and risks at two scales, regional and local.

On the one hand, large scale values and risks associated to RMAs were identified and mapped, by selecting mapping and aggregating suitable criteria to describe them. Then, a series of maps of values and risks were generated, aggregated and used to support the selection of five priority contexts, representing those RMAs with highest aggregated values and reflecting a significant combination between values and risks. On the other hand, positive and negative landscape elements were identified at local scale for each value and risk defined at broader scale. Visibility maps were further generated at local scale in order to assess the visibility from and to the area (a GIS-based function was used).

The next sections provide more detailed information on the method applied.

4.5.2.1 Large Scale Values

The study identified and mapped three large scale values: ecological, landscape and economic.

The **ecological value** refers to the important role played by RMAs in terms of habitat provision and biodiversity protection. Although RMAs are farming-dependent landscapes, their extensive agriculture has a fundamental role in creating environmental quality and providing a number of ecological functions (e.g. filtering out chemicals and noise, providing connectivity between natural areas, etc.) (Briner et al. 2013). To map the ecological value at regional scale we combined:

1. the elements of the ecological network of the Province of Trento as defined and mapped by the Regional Spatial Plan (scale 1:50,000). It consists of: Protected Areas (i.e. Natura 2000 sites, local reserves, etc.), water resources and related zones designed for protecting them, wildlife areas (i.e. glaciers, inhabited forests, etc.);
2. the proximity to Protected Areas calculated through GIS-based buffering function. It considers the role played by agricultural areas with respect to nature conservation (i.e. species dispersal and genetic exchange, filter out of external noise, etc.) as well as the benefits that agricultural areas can derived from the proximity of Protected Areas (e.g. source of phytoparasite predators, favourable influence on microclimate, local groundwater recharge, etc.) (Geneletti 2007).

The **landscape value** considers the unique character of RMAs, shaped and maintained by a continuous interaction between human activities and the environment (CoE 2000). Among the most evident expressions of such interaction are the terraced slopes and their singular elements such as drywalls, man-made rainwater systems, etc. These peculiar elements and attributes generally confer a high visual and cultural quality to RMAs, which is a key property of landscape character (CoE 2000). According to the classification of rural areas providing by the

Regional Spatial Plan, 70 % of RMAs total surface is classified as high-quality rural areas due to both the presence of traditional cultivation systems (e.g. viticulture in terraced slopes) and the landscape scenic beauty. The landscape value further refers to both the importance and the state of conservation of peculiar elements such as cultural and heritage assets, natural monuments, etc., contributing to shape rural landscape (Ricci 2003). The landscape value was mapped at regional scale by considering the presence or absence of elements with a high landscape value as described in the Regional Spatial Plan (PAT 2008). They include: historical settlements, sites with a high landscape scenic beauty, geological and geomorphological monuments, and old roman ways.

The **economic value** reflects the importance of RMAs in directly contributing to local economy and indirectly preserving the environmental and landscape quality (i.e. reducing erosion and flooding risk, contributing to regulating the micro-climate, etc.). From an economic perspective, there are many constraints limiting agriculture in RMAs, including: unfavourable location and geomorphology, limited accessibility, limited opportunity to mechanisation, etc. However, these limitations are often traded off by favourable soil and climate conditions enhancing the production of traditional and high quality goods (Freppaz and Agnelli 2008). Moreover, the economic value of RMAs depends upon their profitable localisation—well-known places associated with typical foods or wines may increase the values due to commercial and non-commercial reasons. To map the economic value at regional scale, we assigned an average agricultural value (€/m^2) to different classes of agriculture land use (vineyards, olive groves, crops, etc.) according to their location. The values were estimated from the official Bulletin of the Province of Trento concerning land expropriations, which provides an average economic value of agricultural land per hectare according to the type of cultivation and the localisation of the farm (BUTAA 2011).

4.5.2.2 Large Scale Risks

The study identified and mapped three risks at regional scale: hydro-geological, environmental, anthropic.

The **hydro-geological risk** considers that geological and water related hazards (landslides, rock-falls, flooding, etc.) are among the most risky events for vulnerable population and human activities in mountain regions (Franceschetti and Argenta 2002; WWF 2012). On the one hand, RMAs can be more exposed to hydro-geological risks due to their particular morphology. On the other hand, their preservation is fundamental to mitigate those risks—for example, rainwater systems and terraces may play a crucial role in controlling landslides (Fontanari and Patassini 2008). The Water Management Plan of the Province of Trento has defined and mapped this risk through an algorithm classifying it in four classes (very high, high, moderate, low), by considering the probability of the hazard (landslide, debris-flow, avalanche and flooding), the value of land use and the likely vulnerability of people and goods exposed to it. This official map was used to map the hydro-geological risk in this project (scale 1:100,000).

The **environmental risk** considers the probability to have damage due to a general exposure (the pathways between the source of the damage and the affected population or resource) to an environmental hazard (the source of the damage) (EEA 1998). On the one hand, agriculture could be seen as a non-point source of pollution (i.e. agrochemicals, etc.). On the other hand, there may be various sources of hazard for rural landscape. In this study, we identified several pressures which may have an adverse effect on rural landscape at regional scale, including: landfills, sewage treatment plants, highways, and degraded urban areas. Then, we mapped the risk by establishing and calculating a series of sensitive distances from these point sources (distance values were established based on APAB 2008—basic GIS-based functions were used).

The **anthropic risk** reflects the pressure on RMAs and rural landscapes associated with land use changes (i.e. urbanization), tourism and mines. To map this risk at regional scale we combined three criteria:

1. the index of tourism pressure, given by the ratio between the maximum number of tourists achieved in 1 year and the number of residents per municipality. It describes the risk associated with tourism-related pressures, including: the increase of traffic and waste production; the greater use of water and energy; etc. It may vary strongly with season and location (winter ski, lakes, etc.). We mapped it by assigning the value of the index to each municipality (PAT 2012);
2. the trend of the Utilised Agricultural Area per municipality, given the change of cultivated land during 18 years (from 1982 to 2000). It describes the risk associated with the abandonment of rural areas and activities. We mapped it by assigning the value of the index to each municipality (e.g. municipalities presenting negative trend have been considered the most exposed to the abandonment of rural activities) (PAT 2012);
3. the presence and the proximity to mines, reflecting the potential risk associated with both noise pollution and visual impact. Basic GIS-based functions were applied to map them.

4.5.2.3 Priority Contexts

The assessment of values and risks at regional scale (supported by aggregated and non-aggregated values and maps) led to the selection of five priority contexts located in different valleys. They represented those RMAs aggregating highest risks and values and reflecting a significant combination among them.

A brief description of these five priority contexts (identified through their ID code) is following (Box 2).

Box 2: Priority Contexts

RMA(1) VC5 (*Val di Cembra*)—area located on steep slopes (average 47.5 %) with scarce accessibility and limited opportunity to mechanisation. More than 70 % of the total surface is cultivated hosting well-conserved vineyard terraces made up by drywalls.

RMA(2) AG2 (*Alto Garda e Ledro*)—area with high landscape and cultural values due to the presence of ancient olive groves stretched on terraces located in both private and public land (about 50 % of the total surface is occupied by olive groves). It has a high visibility (from and to the area). Main risks are associated with: urban development, abandonment and degradation of steepest sectors (stone walls, etc.) and tourism (mainly climbers).

RMA(3) RK2 (*Valle Rotaliana-Königsberg*)—area with high scenic beauty due to the great visibility from the surrounding road and train networks and the presence of vineyard traditional cultivation forms (*pergola*) occupying about 75 % of the total surface. The main risk is associated to the abandonment of this singular cultivation form due to its scarce competitiveness on wine market.

RMA(4) AV10 (*Alta Valsugana e Bersntol*)—greatly heterogeneous area in terms of slopes, aspects, land use covers and farming systems (terraced viticulture, apple orchards, soft fruit orchards and grasslands). It is scarcely accessible and exposed to several risks, including: the abandonment (due to the high fragmentation of cultivated plots); visibility impacts (due to the presence of greenhouses and its high visibility from the main surrounding road network).

RMA(5) VN1 (*Val di Non*)—area mainly cultivated in apple orchards (more than 75 % of total surface) with high economic value and quality. It is greatly transformed by mechanised agricultural practices with several remnant areas occupied by vineyards and forests. The main risk is associated with additional transformations, which may cause: the loss of remnants (semi-natural areas), the increase of hydro-geological risk, visual impacts (presence of hail nets), etc.

Then, for each context we identified positive and negative landscape elements (field works and expert opinions was the main source of information) and we associated them with values and risks previously defined at regional scale. For example, the presence of a protected area in the surrounding of the RMA was considered a positive element contributing to increase ecological value or the presence of degraded rural buildings in the RMA was considered a negative element contributing to increase anthropic risk, etc. (Fig. 4.3).

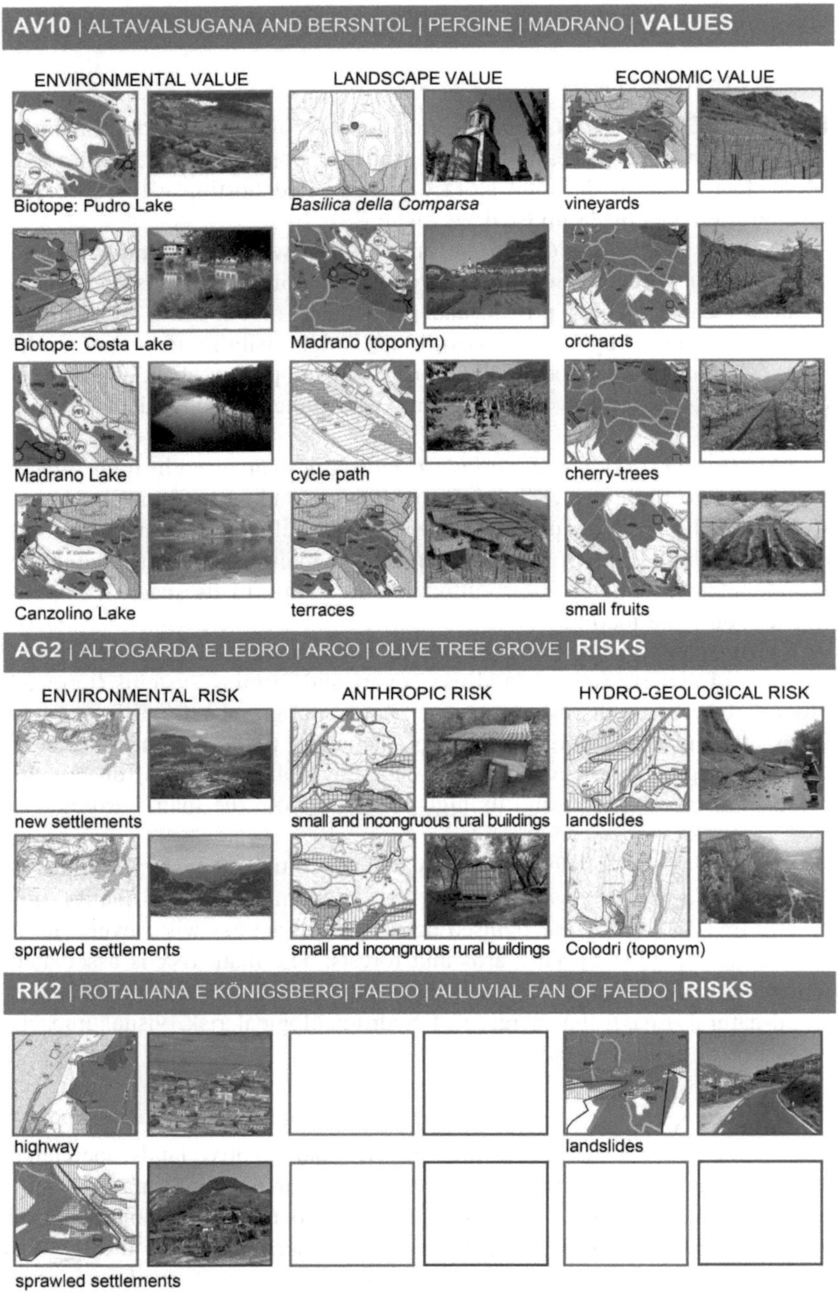

Fig. 4.3 Values and risks in priority contexts. Examples of key landscape elements

This information together with additional data collected during field works was inventoried in a set of graphic sheets, resuming the main features of the five priority contexts, including: dimensional characteristics, morphology, land use properties, visibility (two visibility maps estimating the level of visibility from the area and from the main road network to the area were generated for each context), accessibility, and existing restrictions (e.g. protected areas, protected sites or landscapes, etc.). Moreover, extra information concerning traditional cultivation forms and their changes (e.g. seasonal variations of crops) was reported and simulated through graphic representations.

4.5.3 Definition of Integrated Measures

The five priority contexts were used as case studies where defining and implementing integrated measures to enhance values and mitigate risks previously highlighted. A matrix-based approach was used to support the formulation of these measures.

Firstly, three main strategies for enhancing rural landscapes of RMAs were identified based on the priorities of regional rural policies and spatial plan (see Sect. 4.3.3). They are: protection and conservation, tourism development and environmental sustainability.

Protection and conservation strategy aims to improve environment and rural landscape, by preserving and restoring peculiar rural buildings (drywalls, terraces, rural cottages, etc.), maintaining agro-biodiversity (e.g. preserving and reintroducing endemic seeds and plantations) and promoting related education activities.

Tourism development strategy relies on the multifunctional role of agriculture, aiming at: increasing the competitiveness of the agricultural sector and the value of local products, improving the quality of life in rural areas and encouraging diversification of the rural economy. Among the most relevant interventions are: supporting and incentivising the development of agritourism activities bringing visitors to farms, creating short chain markets for the commercialisation of local products (from earth to table), making land use zoning regulations more flexible.

Environmental sustainability strategy refers to the integration of eco-friendly practices into the production and transportation chains (e.g. using recycled materials for packaging and for replacing non-recyclable materials used in cultivation forms, installation of solar panels for energy production, adopting water-saving irrigation systems, etc.).

Then, three targets for implementing the proposed strategies were identified based on relevant literature (Fontanari and Patassini 2008; Sarzo 2007) and the survey of positive and negative elements in priority contexts. These targets are: rural buildings and assets, production systems, and accessibility. Thus, by crossing strategies with targets, we defined a set of integrated measures aiming to support landscape management in RMAs. Table 4.1 shows a simplified version of the matrix applied and the integrated measures defined.

Table 4.1 Simplified matrix and measures

	Protection	Tourism	Sustainability
A. Buildings and assets	A1 Terraces	A.5 Existing buildings	A.7 Sustainable building and renewable energy
	A2 Existing rural buildings	A.6 Building cubage	
	A3 Small and incongruous rural buildings		
	A4 Cultivation devices		
B. Production systems	B1 Production strategies	B4 Ruaral accommodation	B7 Biodiversity and environmental protection
	B2 Traditional systems	B5 Education programs	B8 Soil protection
	B3 Water regulation	B6 Territorial marketing	B9 Biomasses
			B10 Biological agriculture
			B11 Zero waste
			B12 Water resources
			B13 Short food supply chain
			B14 Runoff water management
C. Accessibility	C1 Principal accessibility	C1 Principal accessibility	C1 Principal accessibility
	C2 Minor accessibility	C2 Minor accessibility	C2 Minor accessibility

Then, a set of interventions was generated for each measure. Two examples are following:

A6—Building Cubage. It represents an incentive to retrofitting and developing rural buildings for agritourism purposes. It was elaborated from the concept of multifunctional agriculture, proposing the possibility to build new volumes for developing agritourism buildings in case there is no chance to renovate the existing ones. The criteria and mechanisms for cubage distribution in new buildings were further defined, including: fulfilment with the geo-morphological restrictions, use of traditional materials, and compliance with visual and observation points;

B7—Biodiversity and environmental protection. It refers to the implementation of environmental interventions in farming areas, including the conservation and reintroduction of endemic types of crops, the conservation and restoration of synergic cultivations (associations of plants), the creation, restoration and management of natural and semi-natural areas which provide crucial habitat for wildlife (riparial zones, hedgerows and trees, natural vegetation remnants, etc.).

Finally, for each priority context only relevant interventions were selected and suggested for protecting, restoring and managing landscape.

4.6 Discussion, Concluding Remarks and Lessons Learned

This chapter aimed to present a new approach to support landscape management in rural mountainside areas of Alps. It was developed within a research project funded by the Province of Trento which financed a series of thematic projects to support regional landscape planning. Accordingly, the overall goal of our project was to provide a decision support tool for landscape planning and management.

Starting from the point that landscape changes occur at different spatial scale and can be managed at different administrative levels, a multi-scale approach combining quantitative and qualitative methods was developed to support the definition of integrated strategies and operative tools for the management of RMAs (*a peculiar rural landscape of Alps*).

The first challenge of the project was to define RMAs since no published definition existed, and then, to create a *census database* of RMAs at regional scale. Although other studies and researches have dealt with rural landscapes characterising by the presence of terraces in alpine regions (Fontanari and Patassini 2008) and in specific alpine Valleys (Sarzo 2007), they do not provide comprehensive analyses and in depth inventories for the case study area, limiting their applicability and effectiveness in supporting landscape planning and decisions. Thus, a *census database* of RMAs was generated by integrating quantitative (spatially explicit criteria) and qualitative (fieldworks and expert opinions) information and analyses (see Sect. 4.5.1). It gave a synthetic description of 214 RMAs identified within the case study region. The estimation of key parameters (i.e. average slope, land use, etc.) from spatial data through basic GIS-based functions has also contributed to provide a systematic description of these areas. Although the accuracy of several data has been affected by the different levels of detail of spatial information (i.e. scale of land use map is 1:10,000 and average scale of detected RMAs is 1:3,000), the census helped to portray the state-of-art of these peculiar landscapes. This is consistent with the ELC, advocating the need to find instruments for assessing the landscape condition and setting up monitoring systems to follow and manage their changes (Antrop 2005).

Starting from these results, the Landscape Observatory of the Province has planned to integrate into the 5-year report on the state of the landscape an indicator to monitoring the abandonment of rural areas taking this census as a reference point to compare future changes. This is particularly significant, since it shows the effectiveness of our analysis for supporting landscape planning at regional scale (future actions could be planned from the results of monitoring) and recognises the importance to preserve RMAs for facing on the abandonment of rural areas.

The first part of the project was further characterised by a stronger collaboration between the project team and external experts and institutions, allowing knowledge and skills to be exchanged (6 meetings were organised during the first five month) and providing key feedbacks to improve the research—particularly useful were the suggestions of local experts on: specific issues such as agronomy, traditional agriculture, etc. and operative measures such as the option to develop a *territorial brand* as an integrated certification of environmental, tourism, and production quality. More sectoral information was provided by the experts of the *Committee of Agricultural Areas*, seeing the research as an opportunity to mitigate the visual and environmental impact of agricultural devices (e.g. coloured hail nets) and farm-related buildings (e.g. fruit storage stocks). Finally, general meetings with other project teams working on different landscape researches financed by the same fund were mainly helpful to share common understanding and integrate preliminary results for the development of a regional landscape vision.

The selection of priority contexts was another crucial task of this project (see Sect. 4.5.2) since it allowed sensitive areas reflecting a complex combination between values and risks to be identified. This is particular important for landscape protection, management and planning, since actions need to be appropriately addressed in priority contexts (CoE 2000). The method applied to assess risks and values was developed from the theoretical framework described in Ricci (2003) which proposes the use of vulnerability assessment and hazard analysis for identifying priority contexts where implementing landscape measures and interventions, and afterwards applied in Bertini et al. (2011). Despite it presented several methodological and operative limitations (e.g. spatial data were not always available, aggregated values were affected by data accuracy, etc.), we believe that this method could be usefully applied to select priority landscapes in other regions. In that case, criteria to describe risks and values should be appropriately redefined and adapted to different contexts. Furthermore, the methodology could be improved by considering the relative importance of criteria composing values/risks, for example through multi-criteria techniques (Gómez-Sal et al. 2003; Geneletti 2007). This would be preferred when specific landscape measures (i.e. restoration of landscape elements) need to be implemented, since a greater relevance could be assigned to one or more criteria (i.e. presence of elements with high quality value) composing the preferred value or risk (i.e. landscape value).

The last challenging task of the project was to propose operative and integrated measures to enhance values and mitigate risks within the five priority contexts (RMAs). This was done by using a matrix-based approach (see Sect. 4.5.3), showing a straightforward tool to translate broader scale strategies (e.g. objective of the RDP) into operative instruments.

However, despite our project has provided evidence-based results and a suitable approach to support regional landscape planning at different scales, its integration into planning and management instruments is still poor. The reason seems to mainly lie on the scarce coordination among different Offices of the Authority (*in primis*: Agriculture and Regional and Town Planning and Landscape Protection) as well as a poor integration among different administrative levels (Province, Communities, Municipalities).

According to the integration principle of the ELC, the new programming period of the EU rural development policy would have to be an opportunity to integrate the results of landscape studies financed by the Province into its new RDP. However, this seems to have been not the case. And referring to the results of our project, this sounds even more remarkable, being RMAs among the most disadvantageous and sensitive rural contexts of the Province, and thereby, among the most in need of rural incentives. We believe that the results of our project could improve the RDP. Among others, the integrated measures defined for enhancing landscape management in RMAs included the definition of mechanisms and incentives which can be easily integrated into the RDP (see Sect. 4.5.3), contributing to better achieve rural development objectives.

However, rural development is still a sectoral matter within the organisation chart of the Province of Trento (as well as in other contexts). For example, the only authority in charge for assigning and evaluating RDP incentives is the Office of Agriculture, although the impacts of rural incentives on spatial patterns and landscape characters are significant. Moreover, since the beginning of the process financing this project, the Office of Agriculture has not been involved at all (neither during the call's preparation nor in the evaluation of landscape research proposals). This lack was perceived since the first meetings organised during the research period when topic (RMAs) and objectives were defined and agreed.

Another opportunity to convert the results of our project into operative planning tools is represented by the forthcoming preparation of 16 spatial plans at Community scale (*Piani Territoriali di Comunità—PTC*). According to the new urban act, defining relevant landscape elements and characters and downscaling those depicted at regional scale by the Regional Spatial Plan (which include high quality rural areas) is among the tasks of the PTC. We believe that the *census database* of RMAs as well as the method applied to select priority contexts and related information generated by this project (e.g. land use cover information, presence of landscape elements and conservation status, visibility, etc.) could provided a helpful tool to support this definition.

Finally, we consider that our project reflects the main principles of the ELC (and in particular the integration) since it deals with different landscape issues at different scales and levels of intervention, proposing conservation and development measures that could be integrated into spatial and sectoral planning instruments and tools. Moreover, we believe that methods used and developed here may be applied in and tested to other social, political and economic scales and contexts for supporting landscape decisions.

Nevertheless, our project suggested that integrating landscape issues and analyses into policies, plans and projects is not enough, since to be effective it seems more important that these measures and tools be undertaken with active cooperation among authorities and related offices, academics, technical and local experts and public. To facilitate this cooperation a political willingness is imperative since landscape policies cannot be separated from rural development or other sectoral policies as the ELC stresses.

Acknowledgments The authors are mainly grateful to Maddalena Ferretti for coordinating the project team with dedication and patience. Special thanks to: Alberto Gelmetti, Francesco Pontalti, Mosé Ricci and Filippo Spaini. And many thanks to: Arch. Nicla D'Aquilio for her kindness; Edmund Mach Foundation and its technical experts for precious information; and the Servizio Urbanistica e Tutela del Paesaggio for providing useful spatial data.

References

Antrop M (2005) Why landscapes of the past are important for the future. Landscape Urban Plan 70:21–34

APAB (Agenzia Provinciale per l'Ambiente di Bolzano) (2008) Modelli matematici per la simulazione di processi di formazione, trasporto e rimozione del particolato. CISMA, Bolzano (Italy)

Bertini A et al (2011) La carta del rischio per i beni culturali: uno strumento per la tutela e la valorizzazione. Paper presented at the XXXII Conferenza scientifica annuale AISRe, Politecnico di Torino, Turin, 15–17 Sept 2011

Bonoldi A, Cau M (2011) Il territorio trentino nella storia europea. L'età contemporanea. FBK Press, Trento (Italy)

Briner S, Huber R, Bebi P, Elkin C, Schmatz DR, Grêt-Regamey A (2013) Trade-offs between ecosystem services in a mountain region. Ecol Soc 18(3):35

BUTAA (Bollettino Ufficiale della regione autonoma Trentino-Alto Adige) (2011) Decisioni della Commissione Provinciale per le espropriazioni di cui all'articolo 3 della legge provinciale 19 febbraio 1993, n. 6 e s.m. valide per l'anno 2011(valori espressi in euro al mq)

Claval P (2005) Reading the rural landscapes. Landscape Urban Plan 70:9–19

CoE (Council of Europe) (2000) European landscape convention (ELC). Vol European Treaty Series No 176:1. http://www.coe.int/t/dg4/cultureheritage/heritage/Landscape/default_en.asp. Accessed 28 Jan 2014

CoE (Council of Europe) (2008) Recommendation CM/Rec(2008)3 of the committee of ministers to member states on the guidelines for the implementation of the European landscape convention. http://www.coe.int/t/dg4/cultureheritage/heritage/landscape/versionsorientation/default_EN.asp. Accessed 28 Jan 2014

CoE (Council of Europe) (ed) (2006) Landscape and sustainable development: challenges of the European landscape convention. Council of Europe, Strasbourg

Diamantini C (ed) (1996) Gli ambienti insediativi del Trentino e dell'Alto Adige. Università di Trento Dipartimento di Ingegneria Civile e Ambientale, Trento (Italy)

EEA (European Environmental Agency) (1998) Environmental risk assessment—approaches, experiences and information sources. Issue Report n. 4

EU Mountain Farming Protocol (2006) On the implementation of the 1991 Alpine convention in the field of mountain farming. Official Journal of the European Union L271/63 30.9.2006. http://www.alpconv.org/en/convention/protocols/Documents/MountainfarmingProtocolEN.pdf. Accessed 28 Jan 2014

Ferrario V (2012) Agricoltura e paesaggi dolomitici. In: De Fino G (ed) (2012) Agricoltura e passaggio nell'arco alpino, Trento (Italy)

Fontanari E, Patassini D (eds) (2008) Paesaggi terrazzati dell'arco alpino. Esperienze di progetto—Manuale. Marsilio, Venezia (Italy)

Franceschetti G, Argenta C (2002) Le montagne: laboratori per uno sviluppo sostenibile. Cleup, Padova (Italy)

Freppaz M, Agnelli A (2008) Recupero Agronomico/Produttivo—Introduzione. In: Fontanari E, Patassini D (eds) (2008) Paesaggi terrazzati dell'arco alpino. Esperienze di progetto—Manuale. Marsilio, Venezia (Italy)

Geneletti D (2007) An approach based on spatial multicriteria analysis to map the nature conservation value of agricultural land. J Environ Manag 83:228–235

Gómez-Sal A, Belmontes JA, Nicolau JM (2003) Assessing landscape values: a proposal for a multidimensional conceptual model. Ecol Model 168:319–341

Haddaway NR, David Styles D, Andrew S Pullin AS (2013) Environmental impacts of farm land abandonment in high altitude/mountain regions: a systematic map of the evidence. Environ Evidence 2:18-24

Huylenbroeck G, van Duran G (eds) (2004) Multifunctional agriculture a new paradigm for European agriculture and rural development. Ashgate, Aldershot (UK)

INEA (Istituto Nazionale di Economia Agraria) (2012) Trentino Alto Adige. http://www.inea.it/trealtadi/profilo. Accessed 01 Feb 2014

Jongman RHG (2002) Homogenisation and fragmentation of the European landscape: ecological consequences and solutions. Landscape Urban Plan 58:211–221

Kothari A (2006) Community conserved areas. In: Lockwood M, Worboys G, Kothari A (eds) Managing protected areas. A global guide. Earthscan, London (UK), p 549–573

Lodatti L, Patassini D (2008) Progetto ALPTER: un modello di partnership che si apre al mondo. In: Fontanari E, Patassini D (eds) (2008) Paesaggi terrazzati dell'arco alpino. Esperienze di progetto—Manuale. Marsilio, Venezia (Italy)

LP (Legge Provinciale) (2008) n.1 del 4 marzo 2008. Pianificazione urbanistica e governo del territorio

MacDonald D, Crabtree JR, Wiesinger G, Dax T, Stamou N, Fleury P, Gutierrez Lazpita J, Gibon A (2000) Agricultural abandonment in mountain areas of Europe: environmental consequences and policy response. J. Environ Manag 59:47–69

MIPAAF (Ministero delle Politiche Agricole Alimentari e Forestali) (2010) Piano Strategico Nazionale per lo Sviluppo Rurale 2007–2013. http://www.reterurale.it/downloads/cd/PSN/Psn_21_06_2010.pdf. Accessed 01 Feb 2014

Morschel F et al (2004) The Alps: a unique natural heritage. A common Vision for the conservation of their Biodiversity. WWF Germany for the WWF European Alpine Programme, Karlsruhe

PAT (Provincia Autonoma di Trento) (2008) Piano Urbanistico Provinciale. http://www.urbanistica.provincia.tn.it/pianificazione/piano_urbanistico_provinciale/normativa_pup2008/. Accessed 01 Feb 2014

PAT (Provincia Autonoma di Trento) (2012) Portale Geocartografico Trentino—Interfaccia Economico Territoriale. http://www.territorio.provincia.tn.it/portal/server.pt/community/interfaccia_economico_territoriale/881/interfaccia_economico_territoriale/37051. Accessed 01 Feb 2014

PAT (Provincia Autonoma di Trento) (2012a) Conoscere il Trentino. Servizio Statistica. http://www.statweb.provincia.tn.it/PubblicazioniHTML/Annuari%20e%20altre%20pubblicazioni%20di%20carattere%20generale/Conoscere%20il%20Trentino/Conoscere%20il%20Trentino%202012.pdf. Accessed 01 Feb 2014

PAT (Provincia Autonoma di Trento) (2012b) Osservatorio Provinciale per il Turismo. http://www.turismo.provincia.tn.it/osservatorio/. Accessed 01 Feb 2014

PAT (Provincia Autonoma di Trento) (2013) Programma di Sviluppo Rurale 2007–2013. http://www.trentinoagricoltura.it/Trentino-Agricoltura/Il-programma-di-sviluppo-rurale-2007-2013/Documenti. Accessed 01 Feb 2014

Plieninger T, Höchtl F, Spek T (2006) Traditional land-use and nature conservation in European rural landscapes. Environ Sci Policy 9:317–321

Pôças I, Cunha M, Marcal ARS, Pereira LS (2011) An evaluation of changes in a mountainous rural landscape of Northeast Portugal using remotely sensed data. Landscape Urban Plan 101:253–261

Priore R (2000) La Convention europénne du paysage. Revue européenne de droit de l'environnement 3:285

Rescia AJ, Pons A, Lomba I, Esteban C, Dover JW (2008) Reformulating the social–ecological system in a cultural rural mountain landscape in the Picos de Europa region (northern Spain). Landscape Urban Plan 88:23–33

Ricci M (ed) (2003) Rischiopaesaggio. Meltemi, Collana Babele, Roma (Italy)

Sarzo A (2007) Il paesaggio dell'abbandono nel circondario agreste di Senter (Valle di Terragnolo, Trentino). Annali del Museo Civico di Rovereto 22:111–170

Sereni E (1961) Storia del paesaggio agrario italiano. Biblioteca universale Laterza, Bari

Spinale and Manez Archive (2002) Spinale e Manez: le Regole—a different way of possessing. Stampa Livio Viviani—Matrix, Tione di Trento (Italy). http://www.regolespinalemanez.it/homepage. Accessed 25 Jan 2014

Tassinari P (ed) (2008) Transformations of landscape in rural territory: reasons of change and future scenarios. Gangemi, Rome (Italy)

WWF Italia (2012) Abitanti silenziosi delle Alpi. http://www.wwf.it/news/pubblicazioni/?uNewsID=1383. Accessed 25 Jan 2014

Chapter 5
The Economic Value of Landscape: An Application for a Rural Area in Northern Italy

Marta Bottero

Abstract From an economic point of view, landscape can be conceptualized as an externality (positive or negative) and landscape values clearly need to be taken into consideration in decision-making process. Among the different methods that can be used for estimating the economic value of landscape, a very important role is played by Conjoint Analysis (CA). CA is a stated preference technique that is it asks respondents to state their preferences and opinions towards hypothetical scenarios. The present chapter investigates the use of the Conjoint Analysis for the estimation of the economic value of a mountain landscape in Northern Italy. Results indicate that people do place a significant monetary value on landscape concepts and the numerical findings could be useful for supporting the definition of future policies for the implementation of landscape schemes.

Keywords Total economic value · Conjoint Analysis · Landscape evaluation · Decisionmaking process

5.1 Introduction

From an economic point of view, landscape can be conceptualised as an externality (positive or negative). This approach was particularly common in the past, when landscapes were seen as incidental co-benefits (or dis-benefits) to activities in economic sectors or product markets. Recently, economists have become more and more aware of the damage done to the environment and landscape by economic activities and they realized that the benefits of these activities should be weighted against the environmental and landscape costs. In this sense, the

M. Bottero (✉)
Interuniversity Department of Regional and Urban Studies and Planning (DIST), Politecnico di Torino, Viale Mattioli 39, 10125 Torino, Italy
e-mail: marta.bottero@polito.it

C. Rega (ed.), *Landscape Planning and Rural Development*, SpringerBriefs in Geography, DOI: 10.1007/978-3-319-05759-0_5,

landscape values clearly need to be taken into consideration in decision-making process (van der Heide and Heijman 2013).

In consideration of this characterization, the use of evaluation tools to estimate the value of landscape can be explained on the basis of two main themes. First and foremost tools are needed to establish and assess the foreseeable benefits of certain actions involving the use and transformation of landscape. Secondly, techniques must be established for the assessment of the effectiveness and efficiency of public expenditure for interventions on landscape. Therefore, landscape assessment can be translated into economic indicators used to design policies for the protection and requalification of landscape (Bottero 2011).

Among the different methods that can be used for estimating the economic value of landscape, a very important role is played by Conjoint Analysis (CA). CA techniques have been widely applied in marketing (Wittink and Cattin 1989), psychology (Green and Srinivasan 1990), transportation research (Hensher 1994) and environmental economics (Adamovicz et al. 1998). A central feature of this approach is that the utility derived from a good or service can be decomposed into "part-worths" relating to different attributes of that good or service (Lancaster 1966). This ability to investigate the "part-worths" of a good is well suited to the nature of environmental goods and landscape. Conducting a CA allows the quantification of values for specific features of a site and the observation of trade-offs people are willing to accept (Kinghorn and Willis 2008; Alvarez-Farizo and Hanley 2002). Moreover, a key element of the CA approach consists in asking individuals about their preferences over different hypothetical alternative scenarios. This is of particular importance in the context of landscape evaluation. In fact, according to the European Landscape Convention, the people's perception of landscape is crucial in the decision-making processes related to the conservation, valorisation and management of landscape (Tagliaferro et al. 2013).

Starting from a real case concerning a rural landscape in Northern Italy, the present chapter investigates the use of the Conjoint Analysis for the estimation of landscape's economic value. In the analysis, the economic value is assessed on the basis of different attributes, such as nature, aesthetic quality, uniqueness and cost. In the evaluation, a questionnaire was developed for the investigation of the preferences of residents and tourists in the studied area with reference to different landscape profiles. The results of the application allow to determine the importance of the different attributes constituting the landscape as well as the economic value of the landscape to be assessed.

5.2 Total Economic Value

From an economic point of view, environmental and landscape elements belong to a particular family of economic goods defined public economic goods. A public good is both non-excludable and non-rivalrous as individuals cannot be effectively excluded from use and where use by one individual does not reduce availability to others. For public goods, such as environment and landscape, market prices either do not exist or only capture a small part of the total value (World Bank 1998). It has been generally agreed that the Total Economic Value (TEV) approach is suitable for dealing with the economic valuation of environmental and landscape goods (Pearce and Turner 1990; Mazzanti 2002).

According to the TEV approach, any good or service is composed of various attributes, some of which are concrete and easily measured, whilst others may be more difficult to quantify. The overall value of the good or service is the sum of all these components. The TEV can be broken down in two main categories of values, namely use value and non-use value.

The use value can be further divided in: (i) direct use value (ii) indirect use value and (iii) option value, whereas the non-use value can be subdivided in: (iv) bequest value and (v) existence value.

Direct use value derives from goods which can be extracted, consumed, or directly enjoyed. As an example, it is possible to consider the evaluation of a forest (Pearce and Turner 1990). In this case, the direct use value would be derived from timber, from harvest of other forest products such as fruit or mushrooms, and from recreational activities, such as fishing or camping in the same forest.

Indirect use value derives from the services the environment provides. For example, wetlands often filter water, improving water quality for downstream users, and natural parks provide opportunities for recreation. These services are referred as ecosystem services, that are the benefits people obtain from ecosystems. The importance of ecosystem services has been recently recognised by several international initiatives, such as the Millennium Ecosystem Assessment (MA 2005) and The Economics of Ecosystems and Biodiversity (TEEB 2010), which focus on drawing attention to the economic benefits of biodiversity including the growing cost of biodiversity loss and ecosystem degradation. According to the aforementioned studies, the economic value of ecosystem services play a crucial role in environmental and landscape decision-making (Rega and Spaziante 2013).

Option value is a special case of use value and corresponds to the value obtained from maintaining the option of taking advantage of something's use value. Existence value derives from the benefits the environment may provide which do not involve using it in any way (for example, it is possible to consider the value people place on the existence of the panda). Finally, bequest value is the value derived from the desire to pass on values to future generations. Figure 5.1 provides a graphical representation of the concept of Total Economic Value.

Different evaluation techniques are available for estimating the TEV. According to the literature (Pearce and Turner 1990), the methods can be divided in two main families:

1. Revealed preferences methods; these methods derive the value of an environmental and landscape resource from the observation of individuals' decisions in real markets. They assume that individuals' affected preferences are estimated through observing the demand for private complementary or substitute goods

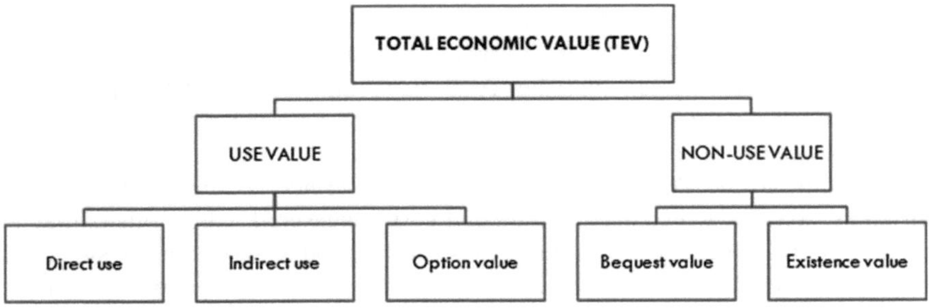

Fig. 5.1 Graphical representation of the concept of Total Economic Value (*source* elaboration from World Bank 1998)

and services somehow linked to the environmental resources. These methods include Travel Cost Method (Voke et al. 2013) and Hedonic Pricing method (Rosen 1974; Freeman 1979).

2. Stated preferences methods: these methods are based on the creation of a simulated market data collection by asking individuals for their opinions or views. In particular, the methods lie on the elicitation of the individuals Willingness To Pay (WTP), that is the willingness to pay of the society for using a certain good or service, or Willingness to Accept (WTA), corresponding to the willingness to accept for abandoning a certain good. Among these methods, it is possible to recall the Contingent Valuation Method (Mitchell and Carson 1989; Carson 2000) and the Conjoint Analysis, which will be described in details in Sect. 5.3.

Figure 5.2 shows the relationship between Total Economic Value and the two aforementioned categories of valuation techniques.

5.3 Methodology

The term Conjoint Analysis refers to a variegated set of mainly statistical methodologies which aim to study individual choices using preferences expressed about various profiles, i.e. several versions of a product or service (Gustafsson et al. 2001; Bravi and Giaccaria 2006; Louviere et al. 2000) and which have in common a number of features (Alvarez-Farizo and Hanley 2002):

1. They are based on a set of attributes or features describing the good, service, project or policy, each taking a number of pre-specified levels.
2. These levels and attributes are combined to build up descriptions of hypothetical bundles, using experimental design techniques.
3. Individuals are asked to state their preferences over these alternatives, using a number of different protocols.

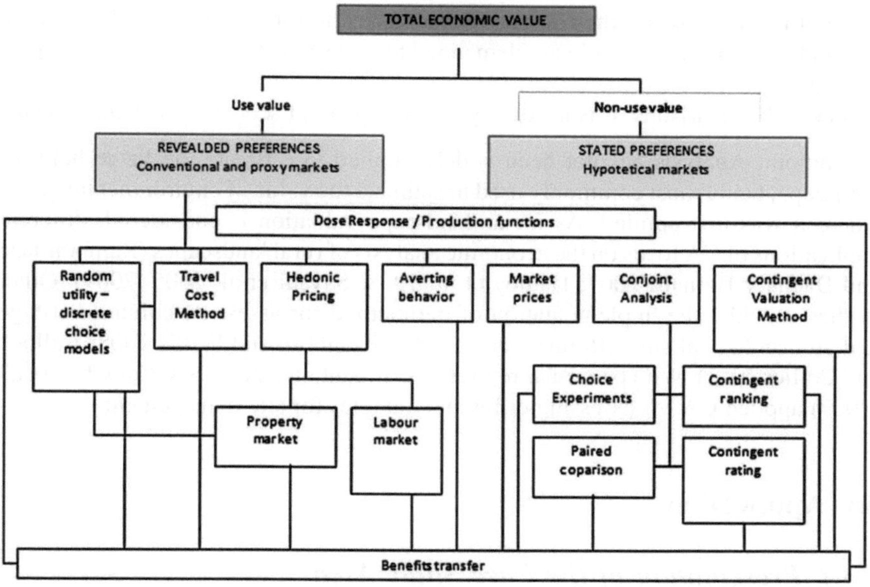

Fig. 5.2 The relationship between Total Economic Value and the two aforementioned categories of valuation techniques

4. During the decision-making process, individuals appraise the worth of each combination, and their choice demonstrates prioritization among the different combinations of features. It is assumed that the total worth of a particular product choice is determined by the different part utilities (part-worths) of each feature level (Sayadi et al. 2005). Responses are then analysed using statistical models.
5. A central feature of this approach is that the utility derived from a good or service can be decomposed into part-worths relating to different attributes of that good or service (Lancaster 1966).

CA is used for analysing the effects of the conjoint action of two or more qualitative features (independent variables) on the preferences of individuals (dependent variables), providing a quantitative measurement of the relative importance of certain features over others. CA is especially suitable for analysing decisions, and in particular when it comes to understanding the process by which consumer/individuals develop their preferences for products or services (Sayadi et al. 2005).

Many variants of the Conjoint Analysis exist, such as Choice Experiments, Contingent Ranking/Rating and paired comparison (Hanley et al. 2001). The existing CA variants can be described as follows:

1. Choice Experiments: it is necessary to choose between two alternatives, versus the status quo;

2. Contingent Ranking: the respondents are asked to rank a series of alternatives;
3. Contingent Rating: people are demanded to rate score alternative scenarios on a scale of 1–10;
4. Paired Comparisons: it is necessary to score pairs of scenarios on similar scale.

Conjoint Analysis has not been widely applied to estimate the value held for landscape, being most commonly used to estimate the value of environmental goods where it was first applied. As far as landscape valuation is considered, different applications of CA focus on the economic analysis of rural landscapes (Rambonilaza and Dachary-Bernard 2007; Hanley et al. 1998; Sayadi et al. 2005, 2009). Other studies consider the implementation of the method for assessing cultural heritage and archaeological sites (Bottazzi et al. 2006; Kinghorn and Willis 2008; Bullock and Collier 2011). With particular reference to mountain landscapes, Campbell et al. (2007) applied CA for assessing different strategies for site management.

5.4 Application

5.4.1 Presentation of the Case Study Area

The Conjoint Analysis approach was applied on a rural landscape in North-West Italy (Bottero 2009). In particular, the area under examination is Susa Valley, a mountain valley very close to the city of Turin in Piedmont Region. The Susa valley has a surface of 1.276,97 km^2 and includes 43 municipalities, with a total population of 116.306 inhabitants.

The case study considers the higher part of the valley featuring many common characteristics of mountain landscapes. In this sense, the High Susa Valley clearly shows the effects of the main trends affecting mountain landscapes in the last years. The decline of traditional farming practice with the growth of tourism activities have been determining a loss of the natural landscape (Fig. 5.3).

The objective of the evaluation is to investigate the importance of the different components of the landscape under examination and to place adequate monetary values on them. The results of the analysis could be used for providing policy recommendations on the welfare gains and losses that society receive from changes in the landscape components (Tagliaferro et al. 2013).

5.4.2 Experimental Design

The first step of the method consists in defining the attributes and the levels of the evaluation model. In this case, five attributes have been identified, namely (i) land use, (ii) openness, (iii) repetitiveness, (iv) naturalness, and (v) cost. Each attribute has been associated to three levels that represent possible options for landscape management.

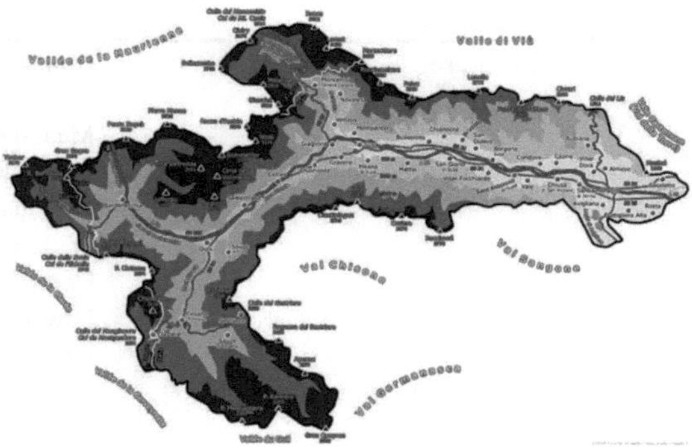

Fig. 5.3 Location of the Susa Valley (Italy)

The attribute "land use" concerns different types of land uses in the area, namely built areas, skiing facilities, meadows and forest. The attribute "openness" is related to amount of space perceivable to the viewer; in this case, three situations are considered, namely presence of visual barriers, punctual vertical elements and wide visual cones. The attribute "repetitiveness" refers to the diffusion of the landscape under investigation; this attribute varies among common landscape, few similar landscapes and unique landscape. The attribute "naturalness" refers to the ecological-natural value of the area, considering the extension of the vegetation; the attributes varies among the levels high, medium and low. The attribute "cost" considers the global expenditure for spending a day in the area, including the cost for internal transports, ski-pass, meals, entrance fees to museums, etc.; this cost varies among 10, 50 and 90 €. In details, the attributes considered for the evaluation are defined in Table 5.1.

It shall be noted that the aforementioned attributes and levels consider the specific characters and values of the landscape under investigation. Moreover, the structuring of the evaluation model has been designed with the help of a specific focus group where the proposed attributes and levels were discussed by several experts in the field of environmental assessment and landscape management in order to reach a common vision. In particular, the Contingent Valuation method was applied within the focus group in order to identify the appropriate bounds and levels for the cost attribute.

Given the number of attributes and levels (Table 5.1), there would have been too many possibilities (i.e., $3^5 = 243$ combinations) to use all them in the survey. In order to identify a minimum efficient set of combinations, an orthogonal fractional experimental design was followed (Addelmann 1962). The generation of the orthogonal design was made using the SPSS software (www.spss.com), where 15 combinations of attributes were selected. Table 5.2 provides the representation of the 15 landscape profiles that have been considered for the development of the analysis.

Table 5.1 Attributes and levels of the evaluation model

Attributes	Levels
Land use	Built areas
	Skiing facilities
	Meadows and forests
Field of view	Presence of visual barriers
	Punctual vertical elements
	Wide visual cones
Repetitiveness	Common landscape
	Few similar landscapes
	Unique landscape
Naturalness	Low
	Medium
	High
Cost	10 €
	50 €
	90 €

5.4.3 Questionnaire

The second step of the model consists in assessing the trade-offs across the landscape attributes identified in the first phase and their marginal value through eliciting people's preferences for the landscape types.

According to the CA methodology, a sample of 30 respondents (tourists and residents in the area) was surveyed with face to face interviews. CA asks people to evaluate several alternatives which are described by different levels of attributes, allowing the attributes to be assessed as well as situational changes (Tagliaferro et al. 2013).

To obtain the information about the respondents' preferences over landscape attributes, a specific questionnaire was developed. In particular, the questionnaire comprised three parts. The first one asked a series of questions regarding peoples' attitudes towards the particular landscape under analysis; the second part concerned the conjoint analysis questions, whilst the third section collected socio-economic information about respondents (age, gender, municipality of residence, job, income etc.).

With reference to the CA questions, the contingent rating approach was followed (Alvarez-Farizo and Hanley 2002). From a methodological point of view, a contingent rating exercise consists of scoring alternative options (hypothetical products, projects, policies or services) using a rating scale. Options are not directly compared with each other, but are evaluated sequentially.

Typically, indirect utility U (as a function of a vector of attributes of the good in question (x) and a vector of socio-economic characteristics (s)) is related to the ratings the individual makes through a transformation function of Eq. (5.1):

Table 5.2 Landscape profiles generated through the orthogonal fractional factorial design

	Land use	Openness	Repetitiveness	Naturalness	Cost (€)
1	Built areas	Visual barriers	Common landscape	Low	90
2	Built areas	Wide visual cones	Common landscape	Medium	10
3	Skiing facilities	Visual barriers	Unique landscape	Low	10
4	Meadows and forests	Punctual vertical elements	Common landscape	Low	90
5	Built areas	Visual barriers	Few similar landscapes	High	90
6	Meadows and forests	Visual barriers	Few similar landscapes	Low	10
7	Meadows and forests	Wide visual cones	Unique landscape	High	90
8	Skiing facilities	Punctual vertical elements	Few similar landscapes	Medium	90
9	Meadows and forests	Visual barriers	Common landscape	Medium	50
10	Skiing facilities	Wide visual cones	Common landscape	Low	90
11	Built areas	Visual barriers	Unique landscape	Medium	90
12	Built areas	Wide visual cones	Few similar landscapes	Low	50
13	Skiing facilities	Visual barriers	Common landscape	High	50
14	Built areas	Punctual vertical elements	Unique landscape	Low	50
15	Built areas	Punctual vertical elements	Common landscape	High	10

$$r_i(x, z, p, s) = \Phi[U_i(x, z, p, s)] \tag{5.1}$$

where r_i is the rating for the good *i*, *z* is a composite good and *p* are prices. Ratings are regressed on the attributes describing the alternatives as in Eq. (5.2)

$$R = \beta_0 + \beta_1 x_1 + \cdots + \beta_k x_k \tag{5.2}$$

In the questionnaire, specific cards were provided to the interviewees in order to ask conjoint questions. In particular, 15 cards representing the 15 landscape profiles considered in the evaluation were given to each respondent. In the card, the profile to be evaluated is defined by a verbal description and a photograph illustrating the landscape under investigation. The respondents were asked to rate each profile on a scale from 0 to 100 (from the least preferred to the most preferred).

As an example, Fig. 5.4 represents one of the cards that were used during the questionnaire.

Profile No. 2

Land use	Built areas
Openess	Wide visual cones
Repetitivness	Common landscape
Naturalness	Medium
Cost	10 €

Score

Fig. 5.4 Example of card used for the evaluation of landscape scenarios

5.5 Results

5.5.1 The Econometric Model

The answers to the CA questionnaire were analysed within the random utility model framework (McFadden 1974). Table 5.3 represents the results of the model.

The parameters t show the effect that each attribute has on the final appreciation of the landscape system under examination. As it is possible to observe from the analysis of the coefficients, the most influent variable on the respondents' preferences is the naturalness (4.176), followed by the price (−3.578), the skiing facilities (−3.192) and the landscape uniqueness (2.444). It is interesting to notice that these coefficient have an high significance as the t distribution is greater than 95 %.

Table 5.3 Results of the Conjoint Analysis model

	Beta coefficients	t	Sig.
(Constant)		13.489	0.000
Built areas	−0.107	−1.917	0.056
Skiing facilities	−0.178	−3.192	0.002
Visual barriers	−0.040	−0.725	0.469
Punctual elements	0.020	0.352	0.725
Unique landscape	0.122	2.444	0.015
Similar landscapes	0.026	0.513	0.608
Naturalness	0.198	4.176	0.000
Price	−0.167	−3.578	0.000

Other important observations comes from the analysis of the overall set of coefficients. The results are consistent with similar findings in the scientific literature (Tagliaferro et al. 2013; Marangon and Tempesta 2008) and show that natural landscape are very much appreciated (the *t*-statistic is 4.176) while the built areas are not preferred (the *t* coefficient of the attribute has a negative sign and it is equal to −1.917). Moreover, respondents did not appreciated the presence of skiing facilities in the landscape (the *t* coefficients is −3.192). In this case as well, the presence of a negative sign highlights that people tend to not appreciate artificial and anthropic elements in the landscape, especially in the case of mountain areas. Following this reasoning, it is possible to state that all the variables related to the mountain and natural environmental are appreciated and so they have a positive sign while the variables related to the urban and human realm are not appreciated and so they have a negative sign.

Finally, it is interesting to observe that, as it was expected, the variable related to the price has an high influence and is negatively appreciated by the respondents (the sign of the variable is negative).

5.5.2 Implicit Prices

The analysis of the data collected with the CA survey can be useful to assess how much individuals value landscape concepts. For this purpose, the estimation coefficients of Table 5.3 are used to evaluate the degree to which the interviewees do trade-offs among the attributes. In particular, from the parameters of the model it is possible to calculate the marginal Willingness To Pay (WTP) for each landscape attribute and to calculated the WTP for a specific landscape.

According to Hanemann (1984), the marginal WTP (or implicit price) for each attribute is calculated as minus the ratio between the estimate of the coefficient for the attribute under examination and the monetary coefficient as in Eq. (5.3):

Table 5.4 Implicit prices

Variables	Implicit prices (€)
Built areas	6.41
Skiing facilities	10.66
Visual barriers	2.40
Punctual elements	−1.20
Unique landscape	−7.31
Similar landscapes	−1.56
Naturalness	−11.86

$$WTP = -\frac{\beta_c}{\beta_y} \tag{5.3}$$

where β_c represents the coefficient of any attributes, and β_y represents the coefficient on cost.

Following Alvarez-Farizo and Hanley (2002), the application of the formula (5.1) provides the implicit prices as reported in Table 5.4.

As an example, according to the implicit prices of Table 5.4, the respondents would be willing to pay 11.86 Euro for increasing of one point in the measurement scale the level of naturalness of the landscape area under examination (i.e., from low to medium or from medium to high). As another example, we can consider the element of skiing facilities. According to the implicit price of Table 5.4, people are willing to pay 10,66 € to prevent the construction of skiing facilities in the landscape; in other words, utility decrease of 10,66 € in presence of skiing facilities.

5.6 Discussion

This chapter attempts to measure consumers' preferences for a mountain landscape. Findings could be useful to inform decision-making in the context of landscape planning and management.

Firstly, the outputs of the application can to support the design of landscape conservation policy. According to the findings of the study, natural landscapes in mountain areas constitute an important component of society's utility function. Therefore, policies oriented towards increasing the level of naturalness would have a positive impact on the demand. This result could be useful for the definition of the amount of subsidies to be paid to farmers that adopt conservation measures for the landscape (Colombo et al. 2006). Let us focus more on this issue. The welfare estimate for the attribute related to naturalness is 11,86 € (Table 5.4). As an example, let us consider the case of the municipality of Cesana Torinese that is located in the area under examination. The agricultural land of the municipality is 900 hectares, the population is 1.026 inhabitants and the overall number of tourists per year is 10.507. By aggregating the estimated value for the involved population (tourists and residents) we can state that the social value of increasing the level of naturalness is 11.533 * 11,86 € = 136.781 €, that corresponds to 151,30 € per hectare of agricultural land. The per-hectare welfare estimate thus obtained expresses what society is willing to pay; for this reason, the numerical value can be used as a reference for the definition of the amount of subsidy that could be paid to farmers for the adoption of conservation measures aiming at increasing the level of naturalness of the landscape in order to verify that the costs are not excessive relative to benefits.

Moreover, the use of Conjoint Analysis could offer some recommendations to policy makers for designing interventions that enhance landscape quality (Sayadi et al. 2009). In the present study, respondents evaluated that the presence of skiing facilities in mountain landscape is more negative than the presence of built areas (Table 5.3). This result could be used to support decisions about the typology of actions admitted in specific areas. In our case, planning policies should prevent the construction of new skiing facilities as they create losses in the utility of the involved population.

Notwithstanding the aforementioned advantages and the suitability of CA methods for estimating economic value of landscape, it should not be ignored that some weaknesses exist. The main critical elements are related both to instrumental and to non instrumental biases that can lead to non robust estimated parameters and values. In this sense, it has been noted that CA methods cannot alone provide the definitive answer to any political decision in the context of landscape planning and management (Carson et al. 2002) but the results of CA methods have to be integrated with more data and qualitative analysis in order to overcome the possible limitations.

5.7 Conclusions

The chapter considers an application of the Conjoint Analysis for estimating the economic value of the different components of the Susa Valley rural landscape.

The main findings of the application show that people place monetary values to landscape concepts. In the present study, respondents attributed an higher value to natural and mountain areas and did not appreciate anthropic and artificial landscapes. The results of the study suggest that stated preferences methods, such as Conjoint Analysis, can play important roles in supporting the definition of future policies for the implementation of landscape schemes, as they give information about of much the society values landscape transformation and each specific landscape characteristic. Moreover, this approach can provide interesting data for the definition of levels of subsidies, taxes and other market based instruments that are used in landscape planning.

It is interesting to notice that the present application proposes an innovative method to landscape evaluation, moving away from a purely ecological approach towards a more anthropocentric approach, where the quality of life of present and future generations has a fundamental importance (Mondini 2009).

Another advantage of CA is its problem structuring approach. Following the so-called "alternatives focus thinking", conventional decision making methodologies have always concentrated on the evaluation of readily available alternatives with the implicit assumption that all decision problems are well structured: on the contrary, the CA approach is based on a deep investigation of objectives and criteria to be considered in the decision process. By means of the approach named "values focus thinking", CA allows to identify desirable decision opportunities and create alternatives (Keeney 1992).

From the point of view of the future work, it would be interesting to expand the number of interviews in order to have an higher number of usable questionnaires and to validate the numerical results of the model.

To this end it is possible to state that the proposed method is suitable for dealing with landscape values and cultural, natural and economic factors and it can be employed for supporting decision-making processes in the context of landscape planning and management in different geographic contexts.

References

Adamowicz W, Boxall P, Williams M et al (1998) Stated preference approaches for measuring passive use values. Choice experiments and contingent valuation. Am J Agric Econ 80:64–75

Addelman S (1962) Orthogonal main-effect plans for assymetrical factorial experiments. Technometric 4:21–46

Alvarez-Farizo B, Hanley N (2002) Using conjoint analysis to quantify public preferences over the environmental impacts of wind farms. An example from Spain. Energy Policy 30:107–116

Bottazzi C, Botter M, Mondini G et al (2006) Evaluation of the tourist demand in management plans for UNESCO sites: the case of the Cinque Terre Park (Italy). In: Environment Identities and Mediterranean Area, 2006. ISEIMA '06, p 367–372. doi:10.1109/ISEIMA.2006.345006

Bottero M (2009) Utilizzo della conjoint analysis per la valutazione di un paesaggio montano. In: Bottero M, Mondini G (eds) Valutazione e sostenibilità: piani, programmi, progetti. Celid, Torino, pp 253–258

Bottero M (2011) Assessing the economic aspects of landscape. In: Cassatella C, Peano A (eds) Landscape indicators. Assessing and monitoring landscape quality. Springer, Berlin, pp 167–192

Bravi M, Giaccaria S (2006) La Conjoint Analysis (CA) nelle valutazioni immobiliari. Aestimum 48:39–59

Bullock CH, Collier M (2011) When the public good conflicts with an apparent preference for unsustainable behavior. Ecol Econ 70:971–977

Campbell D, Hutchinson WG, Scarpa R (2007) Using choice experiments to explore the spatial distribution of willingness to pay for rural landscape improvements. In: Working paper in economics 6/07, Department of Economics, University of Waikato, Hamilton

Carson RT (2000) Contingent valuation: a user's guide. Environ Sci Technol 8:1413–1418

Carson RT, Nicholas EF, Norman FM (2002) Contingent valutation: controversies and evidence. Environ Res Econ 19:173–210

Colombo S, Calatrava-Requena J, Hanley N (2006) Analysing the social benefits of soil conservation measures using stated preference methods. Ecol Econ 58:850–861

Freeman A (1979) The benefits of environmental improvement: theory and practice. The Johns Hopkins University, Baltimore
Green PE, Srinivasan V (1990) Conjoint analysis in marketing: new developments with implications for research and practice. J Mark 4:3–19
Gustafsson A, Herrmann A, Huber F (2001) Conjoint measurement: methods and applications. Springer, Berlin
Hanemann MW (1984) Welfare evaluations in contingent valuation experiments with discrete responses. Am J Agric Econ 66:332–341
Hanley N, MacMillan D, Wright RE et al (1998) Contingent valuation versus choice experiments: estimating the benefits of environmentally sensitive areas in Scotland. J Agric Econ 49(1):1–15
Hanley N, Mourato S, Wright R (2001) Choice modelling approaches: a superior alternative for environmental? J Econ Surv 15:433–460
Hensher DA (1994) Stated preference analysis of travel choices: the state of the practice. Transportation 21:107–133
Keeney R (1992) Value focused thinking. A path to creative decision making. Harvard University Press, Cambridge
Kinghorn N, Willis K (2008) Valuing the components of an archaeological site: an application of choice experiment to Vindolanda, Hadrian's wall. J Cult Herit 9:117–124
Lancaster KJ (1966) A new approach to consumer theory. J Polit Econ 2:132–157
Louviere JJ, Hensher DA, Swaitt JD (2000) Stated choice methods. Analysis and application. Cambridge University Press, Cambridge
MA (2005) Millennium ecosystem assessment ecosystems and human well-being: current state and trends assessment. Island Press, Washington
Marangon F, Tempesta T (2008) Proposta di indicatori economici per la valutazione del paesaggio. Estimo e Terrritorio 5:40–55
Mazzanti M (2002) Cultural heritage as multi-dimensional, multi-value and multi-attribute economic good: toward a new framework for economic analysis and valuation. J Socio-Econ 31:529–558
McFadden D (1974) Conditional logit analysis of qualitative choice behavior. In: Zarembka P (ed) Frontiers in econometrics. Academic Press, New York, pp 105–142
Mitchell RC, Carson RI (1989) Using surveys to value public good: the contingent valuation method. Resource for the Future, Washington
Mondini G (2009) Il progetto di sostenibilità. In: Bottero M, Mondini G (eds) Valutazione e sostenibilità: piani, programmi, progetti. Celid, Torino, pp 23–52
Pearce DW, Turner RK (1990) Economics of natural resources and the environment. Johns Hopkins University Press, Baltimore
Rambonilaza M, Dachary-Bernard J (2007) Land-use planning and public preferences: what can we learn from choice experiment method? Landscape Urban Plan 83:318–326
Rega C, Spaziante A (2013) Linking ecosystem services to agri-environmental schemes through SEA: a case study from Northern Italy. Environ Impact Assess Rev 40:47–53
Sayadi S, Gonzalez Roa MC, Calatrava Requena J (2005) Ranking versus scale rating in conjoint analysis: evaluating landscapes in mountainous regions in southeastern Spain. Ecol Econ 55:539–550
Sayadi S, Gonzàlez-Roa MC, Calatrava-Requena J (2009) Public preferences for landscape features: the case of agricultural landscape in mountainous Mediterranean areas. Land Use Policy 26:334–344
Rosen S (1974) Hedonic prices and implicit markets: product differentiation in pure competition. J Polit Econ 62:34–55
Tagliaferro C, Longo A, Van Eetvelde V et al (2013) Landscape economic valuation by integrating landscape ecology into landscape economics. Environ Sci Policy 32:26–36
TEEB (2010) The economics of ecosystems and biodiversity report for business. Executive summary. United Nations Environment Programme and European Commission

Van der Heide CM, Heijman WJM (eds) (2013) The economic value of landscapes. Routledge, Abingdon
Voke M, Fairley I, Willis M et al (2013) Economic evaluation of the recreational value of the coastal environment in a marine renewable deployment area. Ocean Coast Manag 78:77–87
Wittink DR, Cattin PH (1989) Commercial use of conjoint analysis: an update. J Mark 53:91–96
World Bank (1998) Economic analysis and environmental assessment. Environmental Assessment Sourcebook Update, 23

Chapter 6
Everyday People Evaluating Everyday Landscapes: A Participatory Application of Landscape Character Assessment to Peri-urban Countryside

Rinaldo Grittani, Alessandro Bonifazi and Andrea Tassinari

Abstract This work concerns an application of the Landscape Character Assessment (LChA) method to peri-urban countryside in the metropolitan area of Bari, Italy. Motivations are rooted in an interest for participatory approaches to ordinary landscape research and policy making. Following the principles laid down in the European Landscape Convention, the research design included, beyond desk studies and field surveys, also semi-structured interviews and focus groups to involve local stakeholders. Findings point to a clear potential for LChA to help address issues of scale (despite peri-urbanity being an elusive subject), for interactive action-research methods to harness the growing interest in landscape policy among a diversity of social actors, and for photography to assist the development of alternative approaches to landscape quality. Finally, neo-rural practices emerge as a possible means to reconcile rural development programmes and landscape policy, with a view to coping with the in-betweenness of peri-urban landscapes.

Keywords Landscape Character Assessment · Public participation · Peri-urban social practices · Photography · Italy

R. Grittani (✉)
Agricultural and Environmental Science Department, University of Bari "Aldo Moro", Bari, Italy
e-mail: rinaldo.grittani@uniba.it

A. Bonifazi
Independent Researcher, Bari, Italy
e-mail: alebonifazi@gmail.com

A. Tassinari
Theorema S.A.S., Venice, Italy
e-mail: andrea.tassinari@me.com

C. Rega (ed.), *Landscape Planning and Rural Development*,
SpringerBriefs in Geography, DOI: 10.1007/978-3-319-05759-0_6,

6.1 Introduction

Rural landscapes are gaining momentum in both spatial planning agendas and research communities. These once marginal spaces—when compared to either urban or natural contexts—are more and more being considered as strategic with respect to sustainable territorial development.

In this perspective, peri-urbanization is possibly among the key phenomena driving such a paradigm shift: the blurring of borders dividing cities from the countryside; the emergence of in-between landscapes which show at the same time rural and urban characters, in both physical and socio-economical terms; the intermingling of cultural mindsets that had long remained isolated; all these trends contribute to dissolving the confidence of traditional approaches not only to planning and design, but even to analyzing and describing. Along parallel trajectories, both the target itself (rural landscapes, and peri-urban areas in particular) and the way to re-conceptualize it, are changing at a fast pace.

Under the European Landscape Convention (ELC), landscape means "an area, as perceived by people, whose character is the result of the action and interaction of natural and/or human factor*s*" (CoE 2000a, art. 1). Landscape is therefore recognized as "an essential component of people's surroundings, an expression of the diversity of their shared cultural and natural heritage, and a foundation of their identity" (CoE 2000a, art. 5, letter a). Such newly founded understanding is all-encompassing, since it "covers natural, rural, urban and peri-urban areas" and it concerns "landscapes that might be considered outstanding as well as everyday or degraded landscapes" (CoE 2000a, art. 2).

Hence, the present chapter is aimed at discussing whether the ELC is likely to trigger a renewed approach to landscape policies, spatial planning and rural development, which might prove more sensitive to the distinctive features of peri-urban areas. Here—under either the "fringe" or the "sprawl" hypotheses (Meeus and Gulinck 2008)—the need for widespread landscape awareness in public authorities and civil society is possibly more acute.

The underlying research work proceeded through two stages: as a first step, we adapted the Landscape Character Assessment (LChA) method to test a procedural framework that seemed flexible enough to accommodate significant innovations; secondly, we deepened two key aspects—respectively the use of photography in visual analysis and the role of territorial practices in focusing the links between landscape policy and rural development.

Accordingly, the chapter is divided into five sections. After this brief introduction, we provide a conceptual background to the study. Section 6.3 is about research methodology, while the following Sect. 6.4 illustrates the case-study concerning an application of LChA to a peri-urban area, the town of Valenzano in Apulia Region, Southern Italy. After a discussion of research issues in light of the empirical work and the chosen methodological approach, the chapter ends with concluding remarks covering shortcomings and prospects for further developments.

6.2 Landscape Outside the Box: Paradigm Shifts in Policy Concepts and Practice Following the European Landscape Convention

Ever since it was adopted in Florence (Italy), on 20 October 2000—and all the more so after it came into force on 1 March 2004—the ELC has proved a powerful driver for innovations in public discourses and social practice concerning landscape issues. The Parties to the ELC rejected the view of landscape as "a given assemblage of physical objects, which can be objectively analyzed by the natural or social scientist" and maintained that "it is rather a creature of changeable cultural perceptions and identities" (Olwig 2007, p. 581). Moreover, the ELC understood specific landscape quality objectives as "the formulation by the competent public authorities of the aspirations of the public with regard to the landscape features of their surroundings" (CoE 2000a, art. 1). Based on such principles, the ELC calls for a drastic rethinking of both landscape theory and practice. In the following sub-sections, we address some prominent issues, including: public participation in landscape policies and the social production of landscape; the role of visual analysis (and photography in particular) in investigating landscape quality in everyday living environments; and peri-urbanization.

6.2.1 Re-thinking the Role of Citizens in Landscape Policies

In the public policy domain, the collapse of the twofold separation of government—outward (from civil society and the economy) and inward (into politics and administration), —left the stage to complex and unstable webs of relations cutting across all abovementioned spheres of human agency (Jessop 1995; Healey 2007). Warren (2009) introduced the concept of "governance-driven democratization" to flag how planning and policy making—once the cradle of technocrats and administrators—are witnessing a "rebirth of strongly democratic ideals, including empowered participation, focused deliberation, and attentiveness to those affected by decisions" (Warren 2009, p. 1). The varying degrees of citizens' involvement, and the actual scope of their influence in the processes they participate in, have long been the focus of scholars' efforts (Stout 2010)—at least since Arnstein (1969) unveiled the ambiguous nature of public participation. In short, most participatory processes tend to fall in the first or, to a lesser extent, in the second of the following categories: informative, consultative, decisional (Green and Hunton-Clarke 2003). The key aspects underpinning such diversity may be identified in the articulation of *voice* and *power*—which stand, respectively, for openness and actual capacity to influence decision making—with mediated public negotiation being perceived by some as the most desirable combination (Forester 2007).

On a parallel track, the very foundations of research methodology (especially in the social domain) were shaken by the advent of what Lewin (1946) named action-research to stress the interdependence and interaction between different elements of a situation, between objective and subjective aspects, between theory and practice, and finally between knowledge and change (Amerio et al. 2000)—because when dealing with certain issues "*research that produces nothing but books will not suffice*" (Lewin 1946, p. 35). In other words, on the one hand action-research aims at yielding a contribution to knowledge, while on the other hand it tries to promote a betterment of human conditions (Jennings 2001).

The two strands may be reconciled under the co-production paradigm, as understood by Jasanoff (2004), according to whom knowledge, policy and citizenship are mutually constitutive and innovations in any domain—including landscape planning—ensue from the interactions among decision-makers, experts and lay people.

The ELC seems to reason along similar lines, by providing for the improvement of landscape knowledge to depend, among other factors, on "taking into account the particular values assigned to them by the interested parties and the population concerned" (CoE 2000a, art. 5). The Explanatory Report attached to the ELC text (CoE 2000b, par. 22–23) moves one step further, when it states that "Official landscape activities can no longer be allowed to be an exclusive field of study or action monopolised by specialist scientific and technical bodies" and, to put it even more bluntly, "Landscape must become a mainstream political concern". The acknowledgement of the implications of the approach to landscape policy chosen in the ELC became manifest in the implementation guidelines (CoE 2008, par. II.2.3.A), where it reads that "the concept of landscape proposed by the ELC implies an exercise in democracy whereby differences are accepted, common characteristics found and operational compromises eventually reached; these represent an alternative to the drawing up by experts of hierarchical classifications of landscape qualities".

Turning to practice, it is no surprise that problems in implementing the aforesaid principles abound. Stenseke (2009) points to trust, communication and local influence as vital ingredients in participatory approaches—while warning that the reliance of landscape management officers on epistemological perspectives that hold scientific knowledge as superior to local knowledge may prove a major hindrance. Selman (2004) reviewed the role of a diversity of stakeholders in the management of cultural landscape—from central and local governments, to non-governmental organizations, interest groups and the wider public—providing evidence that the local scale of initiatives may be a key factor in securing effective involvement of community-based organizations. While dwelling further on the interplay between public participation and landscape management, Selman (2006, p. 168) also concluded that everyday landscapes "will be heavily reliant for their stewardship on community engagement, not least because it is local people who are most aware of their values and will have ideas about what ought to be safeguarded or enhanced, and how this might be accomplished".

Landscape planning and management seems to fit well in a growingly participatory approach to public policy making, and emerging practices tend to raise similar issues than in other sectors. However, the links between cultural landscapes—in which physical, cognitive and experiential factors intermingle (Stephenson 2008)—and the social practices that re-produce landscape (West et al. 2006), pave the way to promising research perspectives.

6.2.2 Re-focusing Landscape Quality: Photography and the Discovery of Everyday Living Environments

In order to capitalise on the perception that inhabitants have of their native lands the ELC has brought forth the concept of landscape to the realm of "subjectivity", giving way to important perspectives to photography as a survey instrument of the landscape itself. The relationship between photography and reality has been a key question around which critics, photographers and researchers have confronted and shared for a long period of time. One of the main concerns of photography was that of being autonomous and liberating itself of any bond, "a photographer wants to describe photography and not reality, beyond any theory and boring, well-known misunderstandings of its ability of 'interpretation', because the sole purpose of photographing is photography itself" (Zannier 1991, p. 64). Therefore, photography does not reproduce reality but often generates an absolute disagreement with itself; it has the ability to "recreate a reality that appears to have the traits of ours" (Massimo Cacciari, quoted in Marra 2001, p. 341).

Is it any wonder how photography, viewed as a subjective and complex tool, can return to being useful and support processes that aim to investigate, read and design the landscapes in which we live in? The focus then shifts to photographers: it is they who must make certain that photography will remain as near as possible to reality, so that they may read and interpret the places, reflect on the landscape and the dynamics that move them.

Photography in itself is then neither truth nor falsehood or, perhaps, both one and the other. Because it has been used to provide misleading and distorted information (ex. in advertising), it may also help in providing support for the truth.

As part of the testing of LChA in the case study described in Sect. 6.4, photography was utilized as a research tool for reading and investigating peri-urban landscapes. In the definition of the method, reference was made to the approaches of visual sociology (Faccioli and Losacco 2010; Grady 1999; Parmeggiani 2006; Prosser 1996), to those used in the visual-perceptual evaluations of a landscape (Daniel and Booster 1976; Lothian 1999; Tempesta 2006), but above all to authorial photography that deals with landscapes and architecture (Ghirri 2013; Lugon 2008; Quintavalle 1993; Valtorta 2013; Zannier 1991).

6.2.3 Peri-urbanization and the Evaporation of Urban-Rural Divides

The term "peri-urbanization", first introduced by Bauer and Roux (1976), is associated with both the processes of urbanization that result from the expansion of a city into its neighbouring territories, and the sprawling, which affects rural areas as new settlement patterns become feasible on technological and economic grounds. Both phenomena generate territories where several specific characters of urbanity, such as high population density and the significant extent of built-up areas in relation to the open spaces, coexist with those that are typical of rural landscapes, such as the widespread presence of agricultural businesses and other activities with strictly local markets (Di Mario and Pascale 2009).

Turning to the case-study context, the evolution of settlements and landscapes that have characterized Italy reveals the richness and internal diversity of the peri-urban areas. Until the end of the 1940s, despite being affected by major transformations, the Italian landscape still maintained intact its traditional territorial and settlement structure, with the consolidated city clearly separated from the countryside. In rural areas, the landscape was closely linked to agro-livestock and forestry systems, and about ten distinct agricultural landscapes could be identified throughout the entire peninsula (Sereni 1961). The situation changed radically on the principle of the 1950s; a period from which successive technological, socio-economic and cultural transformations altered the pre-existing territorial structures and generated new settlement and landscaped areas (Turri 1990; Lanzani 2003).

Initially, the process of urbanization only concerned several large cities, but in the 1960s and '70s one witnessed not only a further expansion of cities and their suburbs, but above all the onset of urban sprawl. These phenomena originate from "the triumph of mediocrity and ordinary repetition in the space of a few housing models: the single-family house in the centre of the lot, the warehouse, townhouse, apartment house and shopping mall. A sequence of solitary and amassed presence of buildings without a clear rationale, and without an account of necessity" (Boeri 2011, p. xii). In areas previously used for agriculture, a multitude of isolated and autonomous buildings had risen up. Dynamics that often moved from small towns, rather than major cities, generated a confusing, generic and rambling landscape that "seems to mercilessly reflect on the fragmentation of our society, unbridled individualism that holds no regards toward collective space and fragments the very diverse territories of our peninsula making everything look identical" (Boeri ibid). Hybrid environments are created that blend rural traits with new urban aspects, incorporating industrial and residential-touristic elements, which generate plural heterogeneous landscapes.

One of the main outcomes of this "grand transformation" (Turri 1990) is the emergence of a society in which the values associated with urbanity and rurality tend to get confused: "into an Italy that is relentlessly less and less rural in the classical sense [...] the problems of the system of agricultural production is therefore intertwined with the issues of new forms of urbanity, of safeguarding the environment and landscape, of the new frontiers of social safety nets and of welfare (Di Mario 2008).

Some scholars argue that peripheral urbanity includes geographical areas characterized by the fact that they are no longer pure environments, namely, the compact city or the deep countryside (Mininni 2006), nor fall into an intermediate category between the two, but rather reveal themselves as "another" settlement reality, very heterogeneous and variable internally. This covers the necessity to "reconceptualise" the peri-urban area (Fanfani 2006; Mininni 2006) and the very manner of thinking about the boundary between urban and rural (Lynch 1960; Valentini 2005).

These assumptions, despite the major difficulties of theoretical as well as operational natures, contain aspects of great interest. It would be quite useful to focus on the particularities of peri-urban agriculture (OECD 1979; Heimlich and Anderson 2001); to tailor the manifold functions it may carry out, according to the context where it is practiced (Velazquez 2004); to identify who are the participants who "perform" peri-urban agriculture and who are those who "have an impact" on it (Grittani 2011a).

6.3 Research Methodology

As part of a longer term research project, the present study aims to contribute to addressing the following issues, although with a major focus on issues 1, 2b and 3:

1. putting two innovative tenets of the ELC to the test, in an integrated manner, by applying (a) participatory evaluation approaches to (b) peri-urban everyday landscapes;
2. investigating LChA as: (a) an analytical approach (with its focus on landscape character types and areas) to make collective sense of places at multiple scales (from local to regional); (b) a procedural approach to accommodate the contribution of different actors, both experts and lay people, coming from the government, market and civil society alike;
3. dwelling on photography as an action-research method to grasp perceptual—and visual in particular—aspects of landscape quality;
4. understanding how social practices that relate to neo-rurality and multifunctional landscapes interact with land-use planning, landscape policy, and rural development policies in peri-urban areas.

So far, the research project went through two different stages: at the end of 2011, we applied the LChA method (adapted as discussed in the following Sect. 6.3.1) to a case study-area, and carried out a first round of semi-structured interviews and a focus group (both covered in Sect. 6.3.2). In the second stage, spanning over the end of 2013 and the beginning of 2014, the methodological focus shifted to visual analysis and the use of photography to trigger conversational interactions about landscape quality, policies and practices. The way these activities were designed is explained in Sect. 6.3.3. Semi-structured interviews and a peri-urban walk, followed again by a focus group, were carried out at the second stage (as described in Sect. 6.3.2).

6.3.1 Adapting the Landscape Character Assessment Method

The LChA method, widely adopted in the United Kingdom and other northern-European countries, may be traced back to visual perception-based approaches to landscape analysis and design. However, LChA's conceptual limits and application areas extended over time to encompass conservation policies and spatial planning at multiple geographical scale (SNH and CA 2002). In short, and based on most recent applications, LChA may be described as a procedural framework for landscape analysis and policy support, which integrates desktop studies (including GIS-based spatial analyses), structured assessments provided by human observers moving across the study-areas, and inputs from public participation—with a view to underpinning evaluative judgments.

This research, while taking as a reference the LChA as it was proposed in the manual prepared by the Scottish Natural Heritage (SNH) and the Countryside Agency (CA) for the United Kingdom (SNH and CA 2002), has several new elements, as outlined in Fig. 6.1. The characterization stage began with preparatory activities, relating to the definition of organizational and motivational aspects, and foresaw a desk-based study followed by one on the field (conducted on foot and by bicycle), which was carried out by the authors, based on a data collection protocol adapted after James and Gittins (2007).

The characterization was based on the analysis of the physical aspects of the landscape, of those linked to anthropogenic activities and land use, with its aesthetic and perceptual aspects. Two categories of landscape character constitute the conceptual foundations of the LChA, whose identification varies depending on the scale of analysis:

- *landscape character types* are the result of recurring combinations of geological, topographic, vegetation elements, land uses and methods of settlement;
- *landscape character areas* are geographic locations characterized by a specific identity, the result of socio-territorial dynamics occurring in an original "sense of place", while sharing similar characteristics with other areas, which allow to attribute them to a common type.

The following stage of evaluation considered both the quality and the condition of landscape at the time of the survey, and the land use change processes—on-going and predictable. The evaluation stage and, to a lesser extent, also the characterization stage, have utilized participatory contributions derived from two rounds of semi-structured qualitative interviews and focus groups.

STAGE 1: SCOPING			
Motivations and objectives: 1) No previous application in the area 2) Focus on "everyday landscape" 3) Support planning at municipal level	**Scale of analysis and level of detail**: A single municipality (Valenzano) in the Italian region of Apulia	**People and resources** 1) Three action-researchers 2) Independent, self-funded project	**Degree of stakeholder involvement**: Key players and members of local community involved in all major stages

STAGE 2: DESK STUDY	
NATURAL FACTORS	**SOCIO-CULTURAL FACTORS**
Geology, Morphology, Hydrography, Pedology, Vegetation, Ecology, *etc.*	Land use, settlement patterns, business (industry, crafts, agriculture, service economy), historical and architectural heritage, *etc.*
OUTPUT: Draft map of landscape character types and areas	

STAGE 3: FIELD SURVEY (*semi-structured interviews with stakeholders, photo-elicitation*)			
Structural landscape elements	**Aesthetic and perceptual aspects**	**Perceived landscape character**	**Landscape condition and sensitivity**
Topography, land use, farm size and boundaries, buildings and infrastructure; key geomorphological and vegetation features	View, scale, texture, configuration, soundscapes, smells	Synthesis between structural/perceptual elements and experience of the area (tranquillity, insecurity, isolation, *etc.*)	Accessibility to, and uses of, the area Evidence of ongoing transformation Decay, rehabilitation, *etc.*

STAGE 4: CLASSIFICATION AND DESCRIPTION		
OUTPUT: Landscape characterization		
1) Map of Landscape character types and areas	**2) Description of each area with its distinctive features**	**3) Preliminary assessment of Landscape character condition and intensity**

STAGE 5: EVALUATION DESIGN			
Subjects involved in the evaluation - Action-Researchers - Local experts - Citizens and representatives of civil society organizations	**Subject of evaluation** 1) Character 2) Condition 3) Subjective/socially shared value 4) Ongoing trends and future scenarios	**Methods** 1) **Objectives Matrix** 2) **SWOT analysis** for scenarios and strategies 3) **Focus groups**	**Objectives** 1) **Condition** : Redevelop, Restore, Improve, Protect 2) **Character**: Create/Remove, Recreate/Confine, Strengthen/Mitigate

STAGE 6: EVALUATION (*focus groups*)
OUTPUT: Landscape value judgments and rural development scenarios

LANDSCAPE CHARACTER ASSESSMENT GENERAL OBJECTIVES				
Elaborate landscape management guidelines for major	**Support spatial planning and local sustainable development**	Attach special recognition status to landscape areas	Assist landscape design	Evaluate landscape areas carrying capacity

Fig. 6.1 A flowchart of the LChA methodology. Adapted after SCH and CA (2002, p. 13)

6.3.2 Participatory Approaches to Landscape Research

The involvement of local stakeholders was pursued at each stage, save for the very first step, that is, the definition of the objectives of the study. Two methods were resorted to: qualitative semi-structured interviews and focus groups.

As for semi-structured interviews (Zamunner 1998), they were administered to ten interviewees identified based on the personal acquaintances of the authors, supplemented by suggestions of the respondents themselves. They included key witnesses and ordinary citizens alike (McKernan 1996). The outline of the interview consisted of three sections:

1. perception and knowledge of peri-urban landscapes in the case study area;
2. assessment of the character of the landscape (with respect to both *types* and *areas*);
3. willingness to take action and strategies for the enhancement of landscape quality.

Two focus groups (Morgan 1997; Rodriguez et al. 2011), each involving six participants and organized in the same venue (the seat of a local non-governmental organization), were facilitated by the authors. The group interviews started with the presentation of the project and continued with the discussion of the draft map of landscape character types and areas (with a visual support). The focus groups continued with the attribution of subjective values to each area of character: this task was carried out in a simplified way in 2011—by asking to choose an option on a multiple-choice scale (from "very negative" to "very positive")—whereas in the second research round the participants were confronted with the character/condition matrix presented at stage 4 in the LChA diagram (Fig. 6.1).

With respect to landscape policy scenarios, a semi-structured, multiple choice questionnaire focusing on peri-urban social practices in four landscape character areas (carried out in 2014) replaced a SWOT analysis about a single area (applied in 2011).

Besides the use of photography, illustrated in detail in Sect. 6.3.3, the only other major difference in the methodological approach adopted in the second round consisted in the design of the focus group, which was introduced by a 2-hour-walk across three different landscape character areas (as identified in the draft map). This was done to reconcile two contradictory needs, as highlighted in relevant literature (Propst et al. 2008):

1. group discussions may be facilitated if participants could be collectively confronted with features that contributed to and detracted from rural landscape character;
2. the places in which discussions take place should be free from distractions, neutral, and permit participants to face each other.

6.3.3 Photography in Landscape Analysis and Evaluation

In the present research, photography takes on a different role from that of illustrative, which is generally covered in LChA.[1] Photography has supported the entire process of the study; it has played an important role in the reading, analysis and reflection on the areas that have led to the characterization of the landscape, as well as in the evaluation stage of the landscape.

In the characterization stage, photography had supported the elaboration of profiles for the field survey, for getting acquainted with and studying the areas, by portraying the natural and anthropogenic characteristics, gathering up the uses of the land, characteristic buildings, peculiar elements and atmospheres. In essence, through the use of photographs an attempt was made to capture the character of a landscape.

The process consisted of an interaction between the stage of exploring and photographing the areas of interest, followed by the analysis and selection of the images at the desk. This process has allowed us to gradually "get closer" to the landscape as well as to discover unexpected visions and uncommon places.

The definition of the method has made reference to numerous experiences of photographers who have been confronted with the representation of places (for example, Cresci 1975; Emiliani and Monti 1970; Grittani 2011b; Strand and Zavattini 1955), and has also debated suggestions from the approaches of visual sociology, especially from the experiences related to Grounded Visual Sociology (Faccioli and Losacco 2010; Suchar 1997).

All photographs were taken by one author over three-years period (2011–2013). Although they were selected based on the need to represent all character types and areas (as described in Fig. 6.3), three areas have been covered more extensively because they emerged from the desktop study as more complex and diversified. These are "Ognissanti", "Masseria Marrone" and "Lama Montrone and the inner country". Both panoramic views and details were included, which could trigger discussions on physical, symbolic or experiential aspects.

The approach of visual sociology has been even more relevant in the definition of the methods used to enable a debate with the inhabitants and to understand what their perceptions were of the landscape of the case study area. Photo-elicitation was utilized, that is, a variation of a semi-structured interview that is based on images as well as a number of questions (Beilin 2005; Collier and Collier 1986; Harper 2002). This choice was made because "the use of photos as a stimulus during an interview produces a different interaction between the observer and the observed, in the sense that it can shorten the distances because it does not possess a

[1] Even in the project ECOVAST (European Council for the Village and Small Town), which places a significant emphasis on the use of photography to support the participatory process of identifying the characters of the areas, photography has the sole task of illustrating the character of the landscape (ECOVAST 2006).

strong connotation of the language, and because the focus of communication shifts from the interviewee to photography" (Faccioli and Losacco 2010, p. 111).

Each respondent were shown 40 photographs of the open spaces between the countryside and the city of Valenzano, each asked to select 10 that best represented the local landscape, or at least from which they remained impressed by the most simply because they were familiar places—or quite the contrary, unexpected. The conversation then focused on the selected images in order to understand the relations between the respondents and the places represented, assess the landscape character and discuss observed or potential territorial practices and any possible development scenarios.

6.4 Case-Study: A Landscape Character Assessment of Peri-urban Countryside in Valenzano, Italy

The case-study included in the present chapter is about implementing LChA in a peri-urban area where everyday landscapes are prevalent. These challenging conditions for LChA required many departures from mainstream applications of the method as explained in Sect. 6.3.1.

Brief overviews of the territorial features and policy framework in the case-study area are provided in Sects. 6.4.1 and 6.4.2, whereas results and observations concerning characterization and evaluation are illustrated, respectively, in Sects. 6.4.3 and 6.4.4.

6.4.1 Context: A Brief Overview of the Physical and Socio-Economic Characters of the Case-Study Area

The LChA application concerns Valenzano, a town of about 18,000 inhabitants located in the first belt of minor settlements around the provincial capital city of Bari, in the southeasternmost region of Italy, i.e. Apulia. The municipal territory stretches over 16 km^2; it is predominantly flat except for a few gullies (the so called "Lame") running parallel to each other, along the north-south direction.

As for land use and land cover, forests and semi natural areas are hardly detectable, while the artificial surfaces to agricultural areas (where olive groves largely exceed any other cover) ratio is about 1:5 today—a value that is estimated to double, should all developments foreseen in the proposed municipal urban plan actually take place.

The settlement layout shows a relatively dense urban core, which includes the historical centre; a low-density residential area that developed along the road to Bari; and two isolated, specialized use, sites—providing, respectively, affordable housing and knowledge economy or public administration functions.

6.4.2 Territorial Development Scenarios in Valenzano

The main policy drivers of territorial development in Valenzano appear to fall into either of three categories: the Regional Landscape Plan (RLP), the Rural Development Programme (RDP) and the local development plan.

Under the new Regional Landscape Plan's (RLP)[2] two-tier taxonomy, Valenzano is included in Central Apulia, and more precisely in a basin named after the radial system of gullies converging towards the regional capital (Bari), which lies on the Adriatic coast.

Bari represents the core of a loosely defined metropolitan area, around which a polynucleated settlement pattern is to be found, although the urban fabric tends to agglomerate in linear strips following the main axes of the metropolitan transport infrastructure. When the city of Bari and the first belt of towns around it are taken into account, it appears that agriculture areas tend to be squeezed and eventually replaced or—for what matters most to the purpose of this study—peri-urbanized.

Over the last six decades, as development sprawled, rural areas shrank and became fragmented—eventually turning from dominant to marginal. According to the RLP (Regione Puglia 2014), some gullies disappeared (mostly those with gentle slopes) and the agrarian landscape in general underwent simplification as diversified agriculture land uses and covers were affected either directly (by artificialisation) or indirectly (in terms of growingly untended fields).

In the RLP, Valenzano is described as a transition area in many respects: between peri-urban landscape and fully rural areas—whose main features are a mix of heritage elements (including a medieval church and traditional rural buildings and structures[3]); between coastal plains and the Murgia karstic plateau, which here becomes manifest through its first large step; between olive groves in the northern part of the metropolitan area and vineyards in the southern zone.

The olive tree is currently the predominant crop and occupies 316 ha, which account for 70 % of the utilized agricultural area. The remaining area is occupied by vine, fruit and seed crops (Fig. 6.2a), while unutilized agricultural land represents a significant 10 % of the total agricultural area (Fig. 6.2c, d).

Figure 6.2c is indicative of the difficulties in the agricultural sector, and if read in context with the decline in total agricultural land and the number of farms,[4] as well as their splitting and fragmentation, it illustrates an image of the progressive crisis of agriculture in Valenzano.

---

[2] The plan, whose official name is "Piano Paesaggistico Territoriale Regionale della Puglia" (meaning "Landscape and Spatial Regional Plan of Apulia"), is currently undergoing the public consultation stage prior to final approval.

[3] "Jazzi", that is, dry-stone-walled enclosures to pen in sheep and goats.

[4] According to the National Statistics Office (ISTAT 2014), there were 893 farms in 1970 in Valenzano and they took up 1,326 ha, while in 2010 they had fallen to 327, extending for about 440 ha.

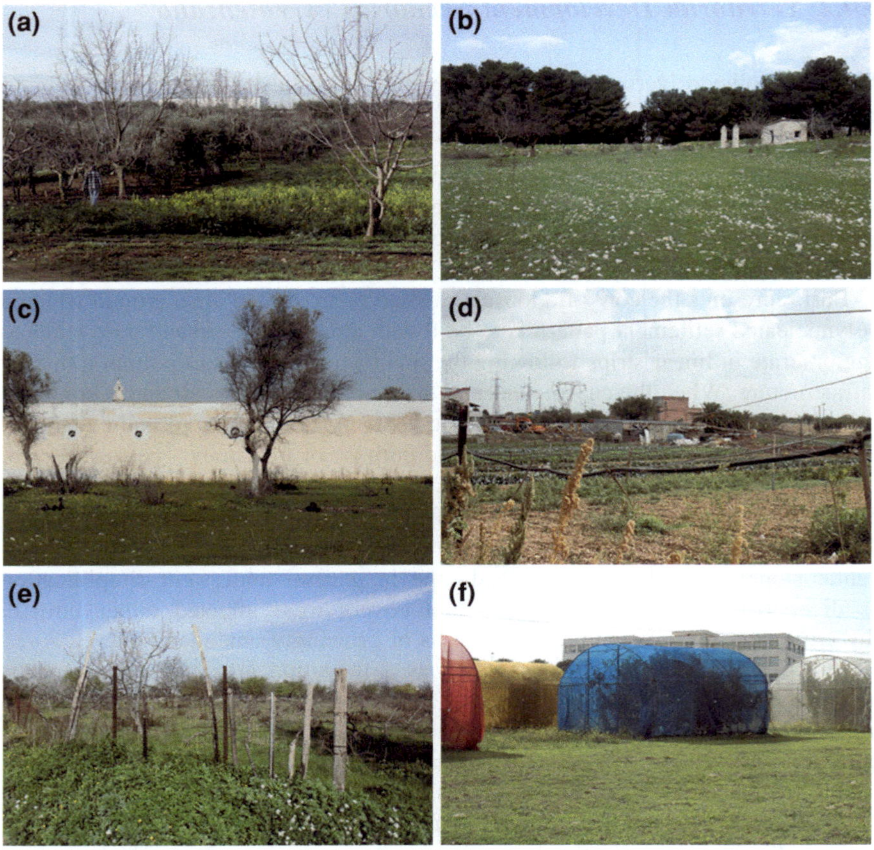

Fig. 6.2 **a** Agricultural activities in Valenzano; **b** Small wood in the "Lama Montrone and the inner country" area; **c** Uncultivated peri-urban area and the wall of a warehouse. In the background, the bell tower of Valenzano; **d** A farm close to one of the main gullies (*Lama Montrone*); **e** An abandoned field; **f** Experimental agriculture in knowledge-based economy landscapes

The image of the border area, from the peri-urban landscape to the rural landscape, is consistently proposed again by the new RLP even within the strategic framework of five projects that do not impose mandatory binding standards, but foretell scenarios to strive for in the medium-long term, possible imageries for a future Apulia landscape.

Within the "City-Countryside Agreement"—the most important project for Valenzano—the northern part of the area is included in both a "multifunctional agricultural park for regeneration" and a "coastal landscape of high natural value"; the southern part of the municipal territory is considered "deep country"; at the turn of these two macro-zones there stands the Inner Countryside, which is of particular interest because it represents the mending bond between the city and the countryside.

The vision of a "multifunctional agricultural park for regeneration", is based on quality agriculture and the protection of agricultural biodiversity on the one hand, and on the other hand, on a series of positive externalities, such as hydrogeological safeguards, conservation of the diversity and connectivity of the ecological networks, and accessibility to rural space.

The "Inner Countryside" plays a vital role in bridging the gap between the city and the countryside—deteriorated by the urban expansion - through an agriculture at the service of its citizens, such as social gardens or suburban parks, the production of fruits and vegetables for local food markets; educational farms, direct harvesting, pet therapy and the like. For Valenzano this could signify the recapturing of fragments of territory by the local community through new territorial practices, the enhancement of the rural landscape and the re-association of urban housing fabrics that are presently detached between themselves.

In Valenzano, an out-dated local development plan dating back to the 1970's, does not address the development of the agricultural territory, as it simply indicates the areas for urban expansion outside of the consolidated city. Hence, spatial local development planning falls short of implementing RLP's strategies.

Within the current RDP (Regione Puglia 2008), the territory of Valenzano falls entirely in the "rural areas with intensive specialized agriculture" type. However, this choice seems to be premised on distance from the other types—"urban agglomerations" and "rural areas with problematic aspects related to development"—rather than on actual correspondence of the observed territorial features to the chosen type. Both the weak analytical capacity—due to a coarse-grained classification at the regional scale—and the lack of a specific policy for peri-urban areas seem to be key to explaining why the 2007–13 RDP has not had any significant impact on a relatively small rural space that is so strictly embedded in a metropolitan area.

The only other spatial public policy that has recently concerned the rural territory of Valenzano is an integrated programme for territorial development[5] in the "Southern terraced ridge of the basin of Bari", proposed and implemented by the municipality of Valenzano, along with two neighbouring local authorities. The programme foresees the enhancement of rural areas, conceiving three itineraries that pass through the urban and agricultural landscape, concerning artefacts and "significant" areas from a landscape and historical and cultural point of view, both in the urban centres and in peri-urban countryside.

[5] In Italian: "Programma Integrato di Sviluppo Territoriale".

6.4.3 Landscape Characterization

At first, the substantial homogeneity of the physical and vegetation aspects in the municipal territory led to envisage just one *type* of landscape character (to be called peri-urban landscape). However, following the examination of in force or proposed policies (including a 2009 municipal spatial planning draft), but above all, based on the outcomes of field surveys and on feedback from participants, we differentiated the four *types* and the numerous *areas of landscape character* identified in Fig. 6.3.

In *multifunctional rural landscapes* there are traces of ecological dynamics, cultural heritage and territorial practices that could be an introduction to development in the multifunctional sense—where agricultural activities take their place alongside agri-environmental measures and recreational functions (Fig. 6.2b).

Productive agricultural areas, mostly identified based on the prospective recovery of currently marginal activity, are defined by both the (relative) integrity of rural settlement patterns and the minor influence of suburbanization, which is otherwise quite evident in Valenzano (Fig. 6.2d).

In *knowledge-based economy landscapes* there is a concentration of supra-local services (universities, national and international research institutes, regional water catchment management authority), which is partly set on the permanence of large estates (Fig. 6.2f).

The built fabric is dominated (in the *city centre*) by residential land uses whose compactness (showing however a very different density) is challenged by the thickening along the main transport routes (*city in progress*) and by the emergence of isolated settlements that are wedged in the agricultural matrix (*urban satellites*). In areas undergoing transformation in particular, LChA captures the development of small industrial estates and the outlooks of urban renewal programmes, re-establishing a representation of the suspension; a landscape of expectation (Fig. 6.4a).

As part of the *multifunctional rural landscapes* character type, one of the areas that indicates the most distinctive features is found in the neighbourhood of the *Church of Ognissanti*[6] (Fig. 6.4c), the main historical and architectural asset of Valenzano, located close to the north-eastern municipal boundary. Although the site has not been valued at all, it is a structure that has marked the history of Valenzano and to which its citizens express a strong sense of belonging. The area of *Ognissanti* is characterized by the presence of rural buildings, more or less recent, spontaneous and vernacular architectures of reduced size and low heights which, while highlighting the need for redevelopment and/or restoration, integrate well in the agricultural context in which they are placed (Fig. 6.4e). For example, one interviewee maintained that "*the area of the Church of Ognissanti has remained agricultural in character despite its potential of urban development, but*

[6] The Church of Ognissanti of Cuti, built in the 11th century, is considered one of the leading examples of Apulian Romanesque architecture.

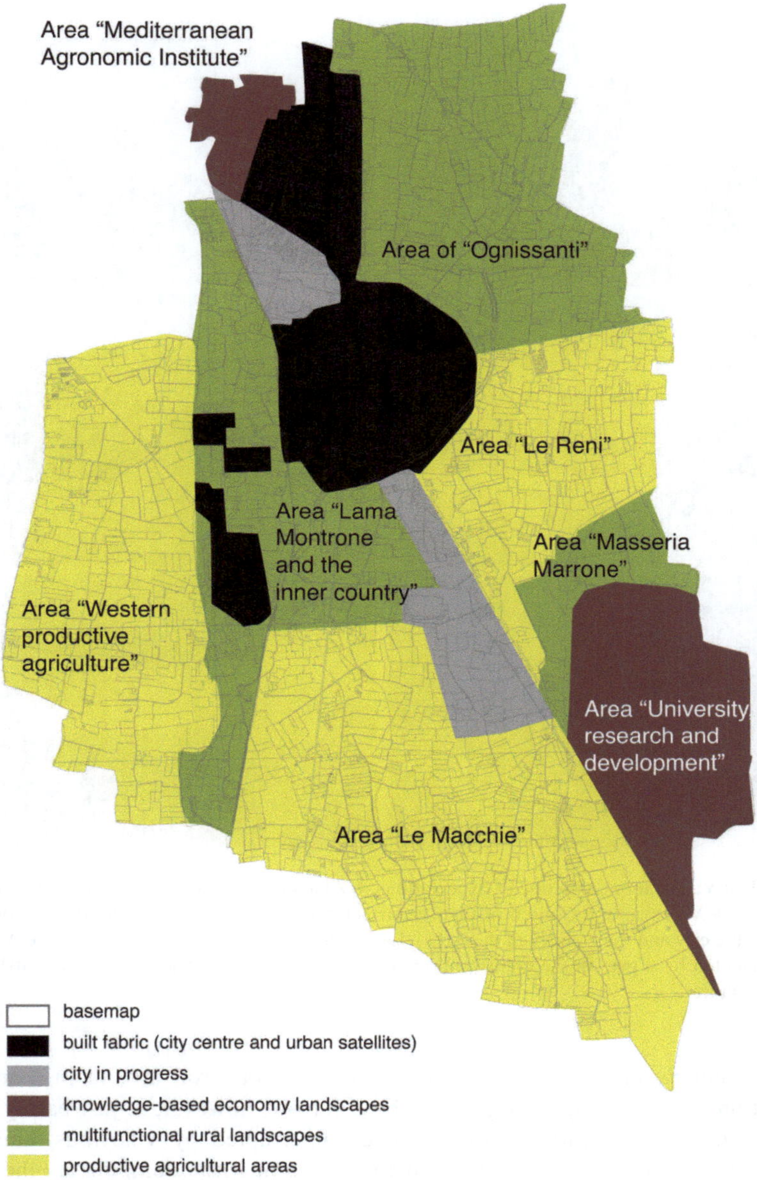

Fig. 6.3 Map of landscape character types and areas

perhaps it is better this way because it is spontaneous, simple and nurtured in an amateur and hobby manner" (resident, researcher).

It is one of the few rural areas visited by not only farmers, retired persons and those who complement their primary work activities with agriculture, but also by

Fig. 6.4 **a** Valenzano: landscape of expectations; **b** Abandoned agricultural landscape in another gully, Lama Valenzano; **c** Church of Ognissanti and an agricultural landscape; **d** The boundary between the city and the country in the Area "Lama Montrone and the inner country"; **e** A plot of land protected by a gate in the Ognissanti area; **f** Former tannery abandoned on the outskirts of the city centre

joggers and *flaneurs* who would pick wild herbs or stroll around in search of relaxation. The crossing of a gully—whose furrow, however, is just barely identifiable in several locations—olive groves, interrupted by vineyards, small or uncultivated gardens, and the occasional presence of waste dumping are the other elements that characterize it. The proximity to the city centre and to a large residential neighbourhood make it an area that, if adequately planned and managed, could also increase its landscape values and capacity to organize stable territorial practices. Several participants have confirmed the peculiarities of this area, pointing out among other things, that the site not only hosted initiatives of a religious nature[7] but also events representing social customs and traditional economics: "*The celebration of November 1 took place at Ognissanti where the "comparizio" occurred. There were stands, many people, the Mass, and we exchanged a red ribbon with the best man*" (resident, theatre actor).

The environmental and cultural elements which led to including the areas "Masseria Marrone" e "Lama Montrone e Inner Country" within the "multi-functional rural landscape" character type are possibly feeble. Conflicting conservation-oriented landscape policy and development-driven spatial planning objectives; the alternation of a widespread decay and some noteworthy natural and cultural heritage assets and; diversified territorial practices, make characterization extremely difficult. Hence, we dwelt in depth on these areas by means of socially interactive research methods. The discussion led to amending the draft map because of convergent observations by several participants who disagreed on certain aspects of the characterization proposed by the researchers.

6.4.4 Landscape Evaluation

The present application of LChA, compared to more popular approaches,[8] has increased the connections between the different stages of characterization and evaluation: bringing forward a sort of evaluation during the field surveys[9] or securing a backwards influence of participatory evaluation on characterization. They are aspects that are likely to increase the potential and usefulness of the method. In addition, the innovations mentioned seem to be essential in order to comply with the fluidity with which participants reformulate their own motivational framework during the course of interactions.

As for capturing subjective value, we observed a higher convergence of judgments—which moreover were globally more positive—in the first group interview. However, several differences in the settings might have contributed to this variance, including: the areas addressed in the two focus groups were different and the second method (i.e. the LChA matrix focusing on character and condition) was manifestly less user-friendly; on the other hand, participants in the second focus group were younger on average, while active citizens were better represented in the first one.

---

7 The fair of November 1 was held in the area of Ognissanti until 1810, the year when it was moved to the city centre. It is noted that the adjoining monastery was first abandoned and then completely disappeared in the second half of the 18th century when the material obtained from the demolition of the abbey walls was used to build a nearby sanctuary.

8 The LChA handbook (SCH and CA 2002) tends to clearly separate the characterization stage from that of the evaluation.

9 The last section of the profile of the field survey, called "Condition and sensitivity of the landscape"—see stage 3 of the LChA flow diagram in Fig. 6.1—can be considered an expeditious anticipation of evaluative aspects.

Table 6.1 A synthesis of the social practices, and the related frequency, associated by participants in the 2014 focus group to each of four landscape character areas

Landscape character areas	Lama Montrone	Le Reni	Masseria Marrone	Santa Croce
Categories of pratices				
Living	3	1	1	3
Working	4	5	5	5
Studying		1	1	
Using	2		4	
Recreating	9	3	6	5
Moving across	5		1	
Sporting	8	5	5	8

Neither the SWOT analysis focusing on landscape quality in a single area nor the multiple choice questionnaire on peri-urban social practices were in fact successful in triggering strategic thinking and landscape policy scenario building, as participants didn't move beyond the mainly descriptive, preliminary stages. The SWOT analysis, carried out for the only area that had elicited full approval ("Ognissanti"), sparked a lively debate on strengths and weaknesses—relating, respectively, to: heritage, accessibility and sense of belonging; weak institutional management capacity. When it came to risks and opportunities, positions and suggestions proved more unclear: for both prospective residential developments in the surroundings and links to the Basilica of Saint Nicholas (in Bari), participants showed different standpoints—verging on conflict. The discussion fell short of putting forward real strategies, as participants tended to focus on single issues rather than making the links between the different factors (e.g. how strengths may help prevent risks). To abbreviate, there was a convergence (albeit with different undertones) concerning the characterization and attribution of value, while the debate on future scenarios has recorded consensus and conflicts.

When participants were confronted with a list of social practices embedded in peri-urban areas—developed by the authors based on the first round of the interactive survey in 2011—and asked to point to those (actually or potentially) more representative of each of four landscape character areas, they seemed to confirm (in one case), and disprove (in another case) the proposed characterization (see Table 6.1). This outcome contributed to a revision of the draft map, with special respect to the inclusion of what had previously been mapped as the northernmost part of the productive agricultural area "Le Macchie" in the multifunctional rural landscape "Lama Montrone and the inner country" (as mirrored in the final version shown in Fig. 6.3).

Social practices in peri-urban areas had been chosen by the authors as a conceptual construct to foster individual reflection and group discussion with a view to making links between personal experience and spatial policies in the case-study area. However, the outcome in terms of strategic thinking and scenario-building was poorly significant, as participants were clearly inclined to speaking about specific places, or to debating general issues more or less in the same terms as they appear to be framed in local public discourses, or rather to engaging in complex narratives.

For instance, the adopted interactive approaches to the evaluation of landscape scenarios have exposed the ambivalence of the concept of spatial development—used by participants in reference to the same area (Ognissanti) to signify either a commitment of public resources for preserving heritage and the sedimented character of the landscape, or the hope to materialize diffused settlement dynamics for residential use. The friction between these alternative views became more evident as the discussion turned to a confrontation on legal and ethical grounds—with special regard to long-standing development rights and recently introduced planning restrictions. Besides the tailored methods used during the focus groups, evaluative arguments emerged also in the course of interviews and the rural walk.

As for interviews, they seem to indicate a shared difficulty that is cast on the peri-urban territory, considered a "repetitive landscape in all respects: morphological, of the cultivations, which also exhibits a significantly greater phenomenon of abandonment" (resident, retired person) if not "a disaster everywhere" (resident, agronomist). Even anyone who has strong emotional ties with these areas displays a strong disappointment for the widespread degradation: "there is no collective and social projection for the rural territory. I had an almond tree…next to which was a plot of land that was a paradise. It is now abandoned" (resident, theatre actor).

The influence of photo-elicitation (used exclusively in the 2013–14 round), seemed to affect the interviewer-interviewee relationship by offering an immediate opportunity for the latter to freely manifest her understanding of, and standpoint on, peri-urban landscape—while by-passing the researchers' frame. Notwithstanding the different approach to interviews, participants' remarks swing between disappointment (for current state of places) and attachment (to heritage and sense of place).

In the words of a resident and entrepreneur: "*Of course the bell tower represents Valenzano, but I associate it to electricity pylons as well*" (Fig. 6.4b); "*These walls strike me, they are the only division from the country: it looks like occupied territories in Palestine, uncertain and temporary situations*" (Fig. 6.4d); "*A gate that doesn't protect anything, it would be nice to leave everything open, but they steal even soil here*" (see Fig. 6.4f). The same participant, however, was surprised to see certain photographs, for they represented places she had not imagined could be part of her everyday landscape.

Another resident (and a physical education teacher) reacted to the view of certain photographs by stating that "*I remember my mother was saddened for all those abandoned fields—she maintained there was a need to set up young farmers' cooperatives to farm the land and fight unemployment*". The same interviewee, then, lingered over a photograph (Fig. 6.4f)—which has been considered representative of the local landscape by others—making unexpected links: "*this building has been abandoned for years, I think it used to be a tannery, it could be turned into a gym—I've been looking for years for a suitable place, but the local authorities didn't show any interest*".

The alternation between familiarity and wonder became manifest during the rural walk, as participants would just go ahead along the route, except stopping in front of unknown buildings (e.g. a huge water reservoir almost nobody was aware of) or unexpected views (a small wood that made two participants wonder: "*How come we never organized scout camps here?*"). While walking, participants came to be exposed to the landscapes they were supposed to discuss about in the focus group, in a way that was likely to emphasize their personal experience of places and make at least some of the relevant social practices visible (in particular, those concerning farming and leisure activities).

6.5 Discussion

The present study, despite the limited scope of organized activities and the preliminary nature of the results obtained, seems to signal several elements of interest. These concern all research issues presented in Sect. 6.3, though to a different extent.

Under the general objective of putting the ELC innovative tenets into practice, we dwelt on the potential role of citizens and diverse social actors in landscape policy. Given a broad conception of landscape is endorsed—one which encompass "the action and interaction of natural and/or human factors" (CoE 2000a, art. 1)—there seems to be a significant societal interest in landscape assessment and policy. More in detail, the main observations one may draw form this study concern the subjects, the methods, and the purpose of public participation. One major shortcoming of this action-research work, so far, has been to limit participants' involvement to individual citizens—mostly highly educated and often relatively active in civil society organizations. Keeping this warning in mind, it is interesting to notice how the boundaries between experts and lay people blurred—as both local experts and knowledgeable citizens would often take the lead during the action-research activities. Moreover, the use of interactive research methods appears to be promising, when not even necessary: the direct experience of peri-urban landscape enabled by the rural walk, and the ensuing situated discussions among participants (Propst et al. 2008), made the activity both more effective research-wise and more empowering for participants themselves. Likewise, photo-elicitation-assisted interviews pushed the relationship between researchers and participants one step further towards the interactive immersion into the subject and the context, that is, local peri-urban landscapes.

There seems to be a clear potential for a contribution of citizens and the local community at large to shaping landscape knowledge and policy in a co-productive manner (Jasanoff 2004). However, such potential inexorably entails getting to grip with conflicting systems of value—as it became evident even based on our very preliminary results. It is our opinion that—once the purpose of action-research were clearly set as influencing landscape policy at both governmental and community self-organization levels—all relevant social actors should be involved in developing creatively integrated visions of sustainable development (Dryzek 2005). Should that not be case, action-research would most likely fall short of facilitating key interactions among those very same social actors, who are entitled on cognitive, ethical and political grounds to re-produce landscape (Magnaghi 2001).

Turning to the specificity of peri-urban landscapes, the literature (Selman 2006) shows that they can prove a very elusive subject for both investigation and policy making. In this framework, it should be emphasized that everyday landscapes (such as peri-urban agricultural areas), characterized by large scale homogeneity, and by extreme fragmentation to that of detail, call for a particularly significant emphasis on experiential approaches and for the active involvement of local communities, as well as for reinforcing qualitative evaluation approaches. Moreover, the concept of landscape character, which is key to the implementation of the LChA method, should perhaps be interpreted in a more problematic way, enriching the connotation in both a positive and negative sense, and questioning some of the assumptions—among which the identification of consistency as a constitutive factor of the character.

When it comes to reflecting on the usefulness of LChA as a methodological approach to landscape action-research projects, an appraisal of the adaptations proposed in this work should take duly into account the inherent limitations of a totally self-funded project that couldn't mobilize neither the interdisciplinary expertise nor the across-the-board involvement of local stakeholders.

With respect to LChA's focus on landscape character types and areas to make collective sense of places at multiple scales, we tested the analytical-descriptive potential of the method, when compared to other spatial analyses that underpinned the elaboration of relevant policies at regional level. Notably, we took care of:

1. reformulating the timing and functions of the different stages in the LChA procedure (see Fig. 6.1), as interviews have provided insights for both characterization and evaluation, and focus groups, while aiming for evaluation and open discussion of policy scenarios, ensured general verification of the proposed characterization;
2. reconceptualising characterization, which was intended to be not exclusively objective (based on desk-study and field survey), open as it was to participants' observations on values and practices, as well as influenced—in a transparent manner—by the most relevant policy frameworks.

Under the aforementioned conditions, landscape characterization seems to offer a significant contribution to making policy scenarios spatially explicit, with special respect to downscaling regional policies and adapting the related zoning instruments to the peculiarities of local contexts. In Fig. 6.5, we compare the way the municipal territory of Valenzano is classified according to two regional policies—the RDP (top left) and the RLP (top right)—with the characterization ensuing from the present work (bottom right). In order to allow direct comparison, a number of adaptations have been made: the agricultural production-oriented landscape types envisaged in the RDP ("rural areas with intensive specialized agriculture") and in the RLP ("deep country") have been subsumed in the LChA's "productive agricultural areas", just the RLP "multifunctional agricultural park for regeneration"

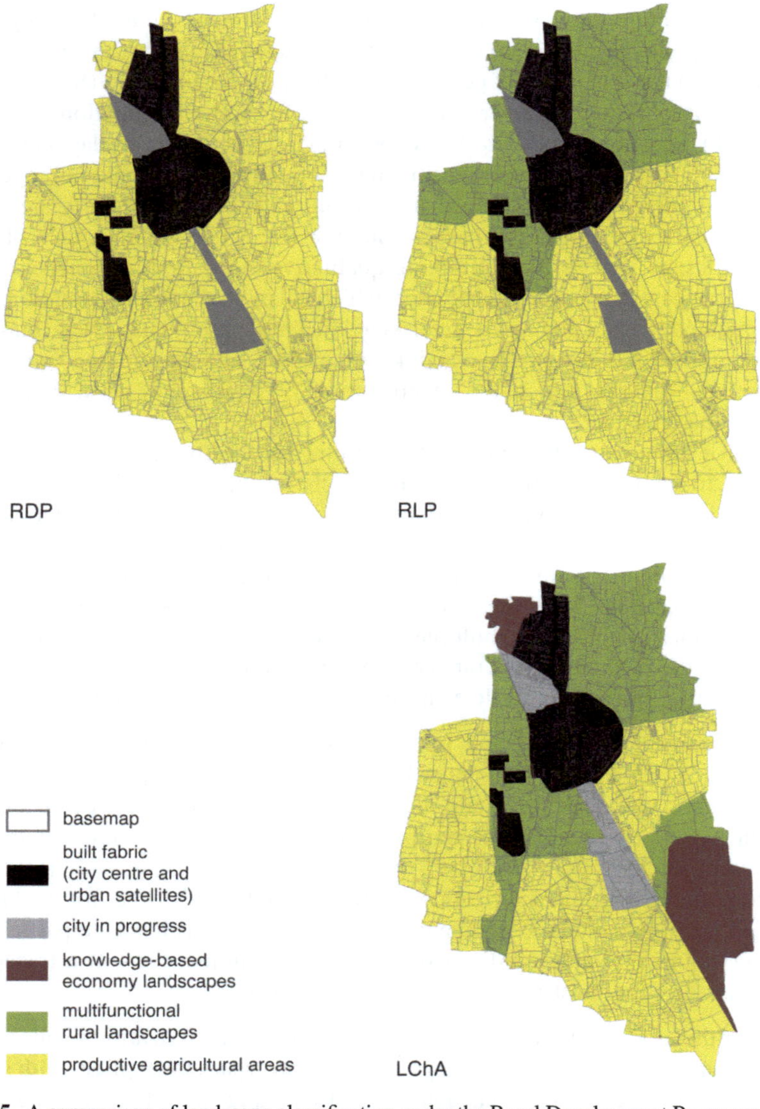

Fig. 6.5 A comparison of landscape classification under the Rural Development Programme (*top left*), the regional landscape plan (*top right*) and the characterization carried out in the context of this study (*bottom right*)

was considered equivalent to LChA's "multifunctional rural landscapes" (see Sect. 6.4.2 for further explanations).

On the one hand, characterization according to the LChA seems capable of complementing the classifications established at regional (RDP) and metropolitan[10] (RLP) scales, by taking duly into account local land uses and related social practices. On the other hand, by tracing back LChA landscape character types to the spatial constructs to be found in regional policy frameworks, local authorities and stakeholders may more easily tap into available resources and cope with existing planning restrictions.

However, zoning proved to be a controversial concept as well, since most participants tended to understand it in the rigidly prescriptive terms conveyed by the existing local development plan—and the main underlying controversies surfaced very easily during landscape characterization-based debates.

The overall evaluation of the support provided by photography to the action-research methodology in grasping perceptual aspects of landscape quality is satisfactory. As explained in Sect. 6.2.2, we didn't aim at supposedly objective uses of photography, but rather positioned our approach at the turn between visual sociology (Grady 1999; Prosser 1996) and authorial landscape photography (Valtorta 2013). In this study, we found with Ghirri (2010, p. 56) that visual representation, just like words and technical information "as an instrument to connect to the world, to establish a relationship with the environment, may have a great cultural importance and a great effectiveness", thus bringing citizens closer to their everyday landscape.

So far, this work involved only researcher-created photographs because, following Prosser (1996), for images to represent empirical findings, researchers need to be photographers as well. These approaches, which are still largely unexplored and strongly unexpressed, appear to be needed in the context of landscape research to activate dialogue and collaboration within the community of residents and users, with a view to elaborate on strategies and development scenarios.

In this study, we understand such strategies in a twofold way. On the one hand, we refer to governmental policies at different administrative levels—with special regard to RDP, RLP and local development planning. On the other hand, it is inherent to action-research that it may result in facilitating participants empowerment in terms of self-governance and bottom-up policy-making processes. In this perspective, it appears that stakeholders and citizens connect more easily to places or issues, although we maintain that territorially-embedded social practices may prove a more effective construct to mediate between everyday life and decision making.

Beyond characterization as zoning, which inexorably evokes the deterministic language of development rights and planning restrictions, participatory landscape evaluation and planning could aim at capturing the tangle of social practices that re-produce peri-urban landscapes. In other words, while investigating neo-rurality and multifunctional peri-urban landscapes, landscape action-research could help fill the gap between strategy-as-practices (Rasche and Chia 2009)—that is, what citizens and other stakeholders in peri-urban communities actually do to reshape

10 For the purpose of the classification shown in Fig. 6.5, the municipal territory of Valenzano is included by the RLP in an area that roughly overlaps with Bari's metropolitan area, and specific planning policies are developed at that scale.

the landscape according to their strategic objectives—and the local, metropolitan or regional policy framework's implicit models of how they are supposed to engender change. In the case-study context, this is made more complicated by the weak support granted by the RDP to landscape quality—not to mention peri-urbanity—while landscape policies and spatial planning, though going through a cycle of radical innovation, have not yet succeeded in establishing alternative conceptions of both development and conservation.

However, we fully subscribe to the idea that "Fragmented peri-urban landscapes will require imaginative strategies which capitalise on their diversity and accessibility, and which create virtue out of their 'in-betweenness'" (Selman 2006, p. 168). Hence, if Valenzano "...needs to reorganize the many 'fragments' which is currently made up of and try to integrate them in a new urban form" (Comune di Valenzano 2010, p. 5), will its peri-urban space help make sense of the social production of the—so far neglected—landscape (Magnaghi 2001)?

6.6 Conclusions

In this chapter, we provided a conceptual elaboration on some innovative principles enshrined in the ELC, with special respect to: (1) bringing local community and stakeholders to the foreground in landscape policy making and; (2) reconsidering landscape quality to encompass everyday landscapes. The investigation is focused on a challenging setting for both tenets: peri-urbanity—which is understood as a hybrid space that blend (neo)rural traits with contemporary urban dynamics—including urban sprawl.

The chapter also reports about a case-study consisting in an application of the LChA method to peri-urban landscapes in a small town located in the metropolitan area of Bari, in Southern Italy. In the mixed-method design, desk studies and field surveys are complemented by interactive social research methods (including semi-structured interviews, rural walks and focus groups). Findings point to a significant societal interest in getting involved in landscape assessment and policy making, and a potential for developing mutually constitutive relationships between citizens and local stakeholders, experts and decision-makers. As for research methods, immersion into the subject, the context and the web of relevant social relationships appear to be both promising and facilitated by the use of interactive approaches—including the direct experience of peri-urban landscape and photo-elicitation.

Landscape characterization as put forward by LChA seems capable of providing a crucial link between the spatial constructs to be found in regional policy frameworks and local land uses and social practices. Finally, the main contribution that the chapter is likely to add to the existing literature lies in strengthening the argument for imaginative strategies to cope with the in-betweenness of peri-urban landscapes (Selman 2006), as well as in bringing neo-rural social practices to the fore as a means to reconcile rural development programmes and landscape policy.

Among the major drawbacks of the present study, it is important to stress the insufficient involvement of selected stakeholders, which were confined to the civil society domain, and the limited scope of the case-study area—two factors that are likely to have emphasised recreational uses to the detriment of productive agriculture. This is however an on-going action-research project, and future developments include, in the medium term, refocusing activities to render outputs more relevant to landscape policy at both governmental and community self-organization levels; testing respondent generated photographs as a further visual research method, and investigating local food market dynamics in neighbouring municipalities.

Acknowledgments The authors wish to thank all the participants in the focus groups and the rural walk, as well as the interviewees—with respect to both the 2011 and 2013–4 action-research activities. We are especially grateful to members of the local branch in Valenzano of the socio-cultural organization ARCI, named "l'Arcipelago", for hosting both focus groups.

Author Contributions

This work stems from an idea of RG, but AB contributed equally to design, literature review, and document analysis. AT joined the other co-authors as from the end of 2013, and has since then contributed equally to the overall research activity. As for conceptual contributions, RG took care of peri-urban landscape studies and the role of photography in qualitative research; AB was responsible for action-research design, with special respect to participatory methods, and the focus on territorially embedded social practices; AT provided spatial analyses and insights into regional landscape policy. With respect to practical work: RG carried out most interviews and field work; AB designed both interviews and focus groups' structure, and facilitated the latter activities together with RG; AT organized the 2014 focus group. To make writing more efficient, Sects. 6.2.2–6.2.3, 6.3.3, and 6.4.3 were drafted by RG, Sects. 6.1, 6.2.1, 6.4.4 and 6.6 by AB, and Sects. 6.4.1–6.4.2 by AT. Sections 6.4.1–6.4.2 were jointly written by RG and AB, as well as Sect. 6.5—which AT also contributed to. All maps are by AT, while all photographs are by RG.

References

Amerio P, De Piccoli N, Miglietta A (2000) La ricerca azione come articolazione di teoria e pratica. In: Amerio P (ed) Psicologia di Comunità. Il Mulino, Bologna

Arnstein SR (1969) A ladder of citizen participation. J Am Inst Plan 35:216–224

Bauer G, Roux JM (1976) La rurbanisation ou la ville éparpillée. Éditions du Seuil, Paris

Beilin R (2005) Photo-elicitation and the agricultural landscape: 'seeing' and 'telling' about farming, community and place. Vis Stud 20(1):56–68

Boeri S (2011) L'anticittà. Editori Laterza, Roma-Bari

Collier J, Collier M (1986) Visual anthropology photography as a research method. University of New Mexico Press, Albuquerque

Comune di Valenzano (2010) Documento preliminare di rigenerazione urbana. Individuazione degli ambiti da riqualificare nel Comune di Valenzano, Relazione
CoE (Council of Europe) (2000a) European landscape convention, Florence, CETS No. 176. Council of Europe, Strasbourg
CoE (Council of Europe) (2000b) European landscape convention, Florence, Explanatory Report, CETS No. 176. Council of Europe, Strasbourg
CoE (Council of Europe) (2008) Recommendation CM/Rec (2008) 3 of the Committee of Ministers to member states on the guidelines for the implementation of the European Landscape Convention. Council of Europe, Strasbourg
Cresci M (1975) Matera. Immagini e documenti. Meta, Matera
Daniel TC, Boster RS (1976) Measuring landscape esthetics: the scenic beauty estimation method. USDA Forest Service, Research Paper RM-167, Rocky Mountain Forest and Range Experiment Station, Forest Service, U.S. Department of Agriculture
Di Mario M (2008) Introductory report at convention "Agricoltura e governo del territorio… trent'anni dopo", INU Lazio—Regione Lazio—Università di Roma TRE, Rome, 29 Sept 2008. http://www.inu.it. Accessed 5 Feb 2010
Di Mario M, Pascale A (2009) Le campagne urbane e le nuove forme dell'abitare. In: Giarè F (ed) Mondi agricoli e rurali. INEA, Roma, pp 23–55
Dryzek JS (2005) The politics of the earth. Oxford University Press, Oxford
ECOVAST—European Council for the Village and Small Town (2006) ECOVAST landscape identification. A guide to good practice. http://www.ecovast.org/papers/good_guid_corr_e.pdf . Accessed 12 March 2010
Emiliani A, Monti P (eds) (1970) Il centro storico di Bologna. Un censimento fotografico. Arte illustrata, Anno III (25–26), pp 97–112
Faccioli P, Losacco G (2010) Nuovo manuale di sociologia visuale. Dall'analogico al digitale. Franco Angeli, Milano
Fanfani D (2006) Il Governo del territorio e del paesaggio rurale nello spazio "terzo" periurbano. Il parco agricolo come strumento di politiche e di progetto. Ri-Vista—Ricerche per la progettazione del paesaggio 4 (6) luglio-dicembre, pp 54–69. http://www.rivista-architettura delpaesaggio.unifi.it/06ri/pdf/06r_fanfani.pdf. Accessed 20 May 2012
Forester J (2007) Public participation as mediated negotiation: entangled promises and practises. Int J Public Particip 1(1). www.iap2.org/displaycommon.cfm?an=1&subarticlenbr=180. Accessed 12 Feb 2014
Ghirri L (2010) Lezioni di fotografia, Quodlibet, Macerata
Ghirri L (2013) Pensare per immagini, Exhibition Catalogue (Rome, 24 April–27 October 2013). In: Fabiani F, Gasparini L, Sergio G (eds) Electa Edizioni, Verona
Grady J (1999) Le potenzialità della sociologia visuale. In: Faccioli P, Harper D (eds) Mondi da vedere. Verso una sociologia più visuale. Franco Angeli, Milano, pp 491–524
Green AO, Hunton-Clarke L (2003) A typology of stakeholder participation for company environmental decision-making. Bus Strategy Environ 12(5):292–299
Grittani R (2011a) Un progetto "aperto" per i paesaggi periurbani. Alcune riflessioni. Territori 3:2–9
Grittani R (2011b) Il Paesaggio di sempre. Upiglio Editore, Milano
Harper D (2002) Talking about pictures: a case of photo-elicitation. Vis Stud 17(1):14–26
Healey P (2007) Urban complexity and spatial strategies: towards a relational planning for our times. Routledge, London
Heimilich RE, Anderson WD (2001) Development at the urban fringe and beyond: impacts on agriculture and rural land. Economic Research Service-USDA, Agricultural Economic Report no. 183
ISTAT (Istituto Nazionale di Statistica) (2014) Italian National Institute of Statistics, agricultural census. http://www.istat.it/it/censimento-agricoltura. Accessed 4 Feb 2014
James P, Gittins JW (2007) Local landscape character assessment: an evaluation of community-led schemes in Cheshire. Landsc Res 32(4):423–442

Jasanoff S (ed) (2004) States of knowledge: the co–production of science and social order. Routledge, London
Jennings G (2001) Tourism research. Wiley, Australia
Jessop B (1995) The regulation approach, governance and post Fordism: alternative perspectives on economic and political change? Econ Soc 24(3):307–333
Lanzani A (2003) I paesaggi italiani. Meltemi editore, Roma
Lewin K (1946) Action research and minority problems. J Soc Issues 2(4):34–46
Lynch K (1960) The image of the city. The Mit Press, Cambridge (Mass.). Italian edition: Lynch K (1964) L'immagine della città (trans: Guarda GC). Marsilio, Padova
Lothian A (1999) Landscape and the philosophy: Is landscape quality inherent in the landscape or in the eye of the beholder? Landsc Urban Plan 44:177–198
Lugon O (2008) Lo stile documentario in fotografia. Da August Sander a Walker Evans 1920–1945, Mondadori Electa, Milano
Magnaghi A (2001) Una metodologia analitica per la progettazione identitaria del territori. In: Magnaghi A (ed) Rappresentare i luoghi. Metodi e tecniche. Alinea Editrice, Firenze
Marra C (2001) Le idee della fotografia. La riflessione teorica dagli anno Sessanta a oggi. Mondadori Bruno, Milano
McKernan J (1996) Curriculum action research. A handbook of methods and resources for the reflective practitioner. Routledge, London
Meeus S, Gulinck H (2008) Semi-urban areas in landscape research: a review. Living Rev Landsc Res 2(3):1–28. http://www.livingreviews.org/lrlr-2008-3. Accessed 29 Jan 2014
Mininni MV (2006) Abitare il territorio e costruire paesaggi, preface to the italian edition of: Donadieu P (2006) Campagne urbane. Una nuova proposta di paesaggio delle città. Donzelli, Roma
Morgan DL (1997) Focus groups as qualitative research, 2nd edn. Sage, Thousand Oaks
OECD (Organisation for Economic Co-operation and Development) (1979) Agriculture in the planning and management of Peri-urban Areas, OECD, Paris
Olwig KR (2007) The practice of landscape "Conventions" and the just landscape: the case of the European landscape convention. Landsc Res 32(5):579–594
Parmeggiani P (2006) Fotografare il territorio: nuovi contributi della sociologia visuale. In: Quaderni del Dipartimento EST, Udine. http://www.mulinococconi.it/pdf/operatori_museali/fotografare_territorio.pdf. Accessed 21 Sept 2012
Propst DB, McDonough MH, Vogt CA, Pynnonen-Valdez DM (2008) Roving focus groups: collecting perceptual landscape data in situ. Int J Qual Methods 7(3):1–14
Prosser J (1996) La costruzione di una metodologia qualitativa basata sulle immagini. In: Faccioli P, Harper D (ed) Mondi da vedere. Verso una sociologia più visuale. Franco Angeli, Milano, pp 39–49
Quintavalle AC (1993) Muri di carta. Fotografia e paesaggio dopo le avanguardie. Electa, Milano
Rasche A, Chia R (2009) Researching strategy practices: a genealogical social theory perspective. Organ Stud 30(7):713–734
Regione Puglia (2008) Programma di Sviluppo Rurale (Rural Development Programme) 2007–2013. http://www.svilupporurale.regione.puglia.it. Accessed 12 Feb 2014
Regione Puglia (2014) Piano Paesaggistico Territoriale Regionale. http://paesaggio.regione.puglia.it. Accessed 12 Feb 2014
Rodriguez KL, Schwartz JL, Lahman MKE, Geist MR (2011) Culturally responsive focus groups: reframing the research experience to focus on participants. Int J Qual Methods 10(4):400–417
SCH (Scottish Natural Heritage), CA (The Countryside Agency) (2002) Landscape character assessment. Guidance for England and Scotland, http://publications.naturalengland.org.uk/publication/2671754?category=31019. Accessed 10 Jan 2010
Selman P (2004) Community participation in the planning and management of cultural landscapes. J Environ Plan Manage 47(3):365–392
Selman P (2006) Planning at the landscape scale. Routledge, London
Sereni E (1961) Storia del paesaggio agrario italiano. Editori Laterza, Roma-Bari

Stenseke M (2009) Local participation in cultural landscape maintenance: lessons from Sweden. Land Use Policy 26:214–223
Stephenson J (2008) The cultural values model: an integrated approach to values in landscapes. Landsc Urban Plan 84:127–139
Stout M (2010) Climbing the ladder of participation: establishing local policies for participatory practice. Public Adm Manag 15(1):46–97
Strand P, Zavattini C (1955) Un paese. Einaudi, Torino
Suchar CS (1997) Grounding visual sociology research in shooting scripts. Qual Sociol 20(1):33–55
Tempesta T (2006) Percezione e valore del paesaggio. Franco Angeli, Milano
Turri E (1990) Semiologia del paesaggio italiano. Longanesi & C, Milano
Valentini A (2005) Progettare paesaggi di limite. Firenze University Press, Firenze
Valtorta R (ed) (2013) Luogo e identità nella fotografia italiana contemporanea. Giulio Einaudi editore, Torino
Velazquez B (2004) Multifunzionalità: definizione, aspetti tecnico-economici e strumenti. In Henke R (ed) Verso il riconoscimento di una agricoltura multifunzionale. Teorie, politiche, strumenti, ESI, Napoli
Warren ME (2009) Governance driven democratization. Crit Policy Stud 3(1):3–13
West P, Igoe J, Brockington D (2006) Parks and peoples: the social impact of protected areas. Annu Rev Anthropol 35(1):251–277
Zammuner VL (1998) Tecniche dell'intervista e del questionario. Il Mulino, Bologna
Zannier I (1991) Breve storia della fotografia d'architettura. Editori Laterza, Roma-Bari

Chapter 7
Tacking Stock: Conclusions and Recommendations

Carlo Rega

Abstract This chapter summarises the main points brought out by the previous contributions to draw key lessons learnt and policy recommendations. These are: (i) the need for multidisciplinary approaches and the mobilisation of a variety of knowledge and expertise, and the consequent need of adequate tools and techniques; (ii) the importance of considering the perceptive component of landscape, in order to capture its distinguishing features with respect to other concepts; (iii) the consequent importance of combining expert opinions and technical tools with methods able to survey people's preferences and opinion, hence the importance of public and stakeholder engagement; (iv) the recognition of the presence of trade offs between different ecosystem/landscape services as well as between rural development objectives and the consequent implications for policy making; (v) the need of a change of paradigm in current approach to landscape and rural development policy design and implementation from a sectoral to a *territorial governance* approach.

Keywords Rural development · Landscape planning · Public and stakeholder engagement · Ecosystem services · Territorial governance

7.1 Lessons Learned and Policy Recomendations

This publication aimed to address both conceptually and operationally the challenges posed by the implementation of two major EU policy objectives towards sustainable development: landscape preservation and enhancement; and rural development. The underlying thesis of the book is that these two domains have until now developed relatively independently from each other, whilst a deeper

C. Rega (✉)
Interuniversity Department of Regional and Urban Studies and Planning (DIST), Politecnico di Torino, Viale Mattioli 39, 10125 Torino, Italy
e-mail: carlo.rega@polito.it; carlo.rega@jrc.ec.europa.eu

C. Rega (ed.), *Landscape Planning and Rural Development*, SpringerBriefs in Geography, DOI: 10.1007/978-3-319-05759-0_7,

integration would generate mutual benefits ultimately contributing to the achievement of their objectives. The policy base to such an assumption, in the EU context, lies in the very foundation of the European Landscape Convention (ELC) and the Rural Development Policy, the second pillar of The CAP. The former explicitly calls for the integration of landscape policies with other sectorial policies at different government levels, and emphasises the role of agricultural activities in shaping and characterising the landscape. The latter identifies (rural) landscape preservation and enhancement as a key priority for the development of rural communities.

Throughout the book, several examples of possible ways to foster the advocated integration were provided. Each chapter presented one or more case studies, as well as tools and methods, with reference to the instruments that implement the ELC and the rural development policy, namely landscape plans and Rural Development Programs. In this concluding chapter, we try to sum up what we deem are the main points emerged to take stock of the lessons learned and turn them into policy recommendations, with a view to the implementation of the ELC and the new rural development policy. The following points, brought out by the different contributions, seems crucial to us:

- The need for multidisciplinary approaches and adequate tools and techniques.
- The importance of the perceptive component as a distinguishing element of landscape.
- The involvement of the public and stakeholders.
- The issue of trade-offs.
- Needed changes in current modes of governance.

The call for multidisciplinary and integrated approaches to deal with complex issues may sound like a *deja-vu* to many scholars or policy makers from different fields. However, the inherently multidimensionality of both the concepts of landscape and rural development shall be acknowledged and stressed. As Lefebrve et al. (2012) put it, of all the environmental public goods provided by farming, landscape is probably the most difficult to describe due to its multidimensional character and the overlap with other public goods. By the same fashion, the rural development policy in the EU is intrinsically multidimensional, as it seeks to achieve, at the same time, three different broad objectives: improving the production and the working conditions of farmers, protecting and enhancing the environment (and landscape) and fostering socio-economic development of rural communities, including non-farmers. This requires that multiple expertises are mobilised, and consequently multiple methods and tools used. In the text, examples have been provided of GIS-based methods (Rega, *infra*, Chap. 2, Cassatella and Seardo, *infra*, Chap. 3; Bragagnolo et al., *infra*, Chap. 4); economic evaluation techniques (Bottero, *infra*, Chap. 5); as well as participatory tools like Landscape Character Assessment (Grittani et al., *infra*, Chap. 6) comprising desktop studies and field works. The first recommendation ensuing from this publication is thus that different methods be used in a combined and complementary way to grasp the many facets of the landscape/RDP interface.

As emphasized by the ELC, landscape is the territory *as perceived by people*. This is a very often quoted definition, but much less often the implications of such a conceptualisation are fully addressed at the operative level. To deal with landscape, hence with landscape preservation in and through RDP, requires that the perceptive component be fully taken into account. As stressed by Cassatella and Seardo (*infra*, Chap. 2), approaches that consider solely the ecological/environmental component of landscape may be misleading. With this premise, specific tools able to catch this dimension are needed. Again, a variety of approaches were presented:

- GIS-based ones like the Visual Sensitivity indicator (Cassatella and Seardo, *infra*, Chap. 2); map overlaying techniques to identify priority areas (Rega, *infra*, Chap. 2; Bragagnolo et al., *infra*, Chap. 4).
- Methods attempting to directly survey people's perception and preferences through the use of photography and questionnaires (Bottero, *infra*, Chap. 5; Grittani et al., *infra*, Chap. 6), or through more involving forms of participatory research, like workshops, interviews and "rural walks" (Grittani et al., *infra*, Chap. 6).

Recognising the importance of the perceptive dimension of landscape has significant consequences also in relation to the objectives of rural development, as effectively pointed out, for instance, by Domon (2011), who argues that the landscape component is a major driving force in the physical and social restructuring of rural areas, as demonstrated by several studies in the past 10–12 years. This body of research shows how visual/aesthetic elements strongly affect residential choices and demographic changes in rural areas (McGranahan 2008; Talandier 2009). Landscape therefore emerges as a fundamental resource for rural development: just like soil quality and presence of other resources (e.g. water) have been key factors for the development of rural areas in the past, "the amenity quality of landscape is destined to act as the main resource for rural areas development in the 21st century" (Domon 2011, p. 339). This in turn means that landscape will increasingly be a discriminating factor for competing territories. Rural areas rich in amenities and offering high visual quality will be the object of strong development pressures, whilst "ordinary landscapes" could be confronted with a new kind of devitalisation (ibid.). The challenge is therefore not only to ensure maintenance of highly valuable landscape, but also to create new ones, by transforming "ordinary landscapes—for example, such landscapes which would be highly marked by intensive agricultural [...]—into landscapes rich in amenities and visual qualities. [...] these areas need to be re-qualified, i.e., in keeping with their particular context, to render them more suitable to aesthetic values but also to environmental, heritage, and identity values" (ibid, p. 339). It has been shown that RDP and agri-environmental schemes (AES) in particularly represent an ideal tool not only to improve the environmental performance of agriculture, but also to enhance the visual attractiveness of ordinary (as peri-urban) landscapes, through the creation of natural and semi-natural vegetation, crop rotation, tree planting and so on (Rega, *infra*, Chap. 2). "Ordinary" landscapes are the main object of the case study presented by Grittani et al. (*infra*, Chap. 6), that highlights both their importance as research field and space of social identity.

The above considerations are linked to the next two points: if landscape is perception, then it is intrinsically a matter of values. But values are not the same for all the people and segments of the society, so this plurality must be acknowledged and dealt with. People do attach an economic values to landscape features, as shown by Bottero (*infra,* Chap. 5), which has important consequences for spatial/landscape planning and RDP, in particular as regards tourism development. Beyond economic aspects, it also clearly emerges how social actors are interested in getting involved in landscape assessment and policymaking, which constitutes a basis to develop mutually constitutive relationships between citizens and local stakeholders, experts and decision-makers (Grittani et al., *infra*, Chap. 6). Stakeholders and public involvement thus appears a key crosscutting activity to reconcile landscape planning and RDP; this represents a challenge for researchers, as meaningful public engagement requires specific expertise, time and financial resources.

This, however, shall not lead to the conclusion that by involving stakeholders and the public, conflicts will disappear. Going back to the concept of ecosystem services (see Rega, *infra*, Chap. 1), it shall be highlighted that not rarely they trade-off against each other. In rural areas, the most typical trade-off is between food provision and regulating/cultural services, but other ones exist: e.g. between renewable energy production (and CO_2 saving) and aesthetics, as in the case of construction of wind farm in valuable landscapes. Other studies have demonstrated that management aimed at biodiversity maximization may undermine other ecosystem services (Chan et al. 2006). No landscape configuration and land use pattern can simultaneously optimize the provision of all types of services (Turner 2010): research should aim to interpret the relation between a certain configuration and the provision of different service, and to show how marginal changes in those configurations affect the flow of such services (Kinzig 2010). This is even more the case for those ecosystem services more directly linked to the landscape and subjective components, as cultural ones. The contribution by Cassatella and Seardo (*infra*, Chap. 3) offers an exemplification of potential trade-offs between a cultural service (aesthetic) and other ecological services in the case of vegetation planting along panoramic roads. Since ecosystem services are by definition related to their utility for people, identifying how do they trade-offs means to identify potential conflicts between different stakeholders and sectors of the society. Indeed, the recognition that trade-offs exist and the consequences that this poses shall be a pivotal aspect of any research dealing with landscape and rural development and is a critical part of decision-making and management (Kinzig 2010).

This leads us to the concluding consideration, perhaps the most relevant for policy and decision-makers. The case studies presented have shown a variety of techniques, tools and approaches that can be effectively used in the field of landscape policies and rural development. Difficulties surely exist, and further researches and application in different geographical contexts are needed. Most of all, as already highlighted, the mobilisation of multiple fields of knowledge and expertise is required: in the case studies, we have seen examples from the fields of spatial/landscape planning, environmental sciences, economics, social sciences and even photography. The contribution by Bragagnolo et al. (*infra* Chap. 4) offers an interesting example of a research project characterised by a strong collaboration between different experts and institutions, allowing knowledge and skills to be exchanged, providing key feedbacks to improve the research and inform policy-making. However, the same chapter also points out that this is a necessary but not sufficient condition for triggering cooperation between different sectors of government and institutions. Again, the necessity of deeper horizontal and vertical integration in government systems to achieve policy objectives in complex contexts is not a novel claim by scholars, but in the case of landscape and rural development this maybe represents the main factor currently hindering their implementation and integration. This is particularly true for the rural development policy that, in the EU context, is still permeated by a sectoral approach—a legacy of its original conception and development as part of the CAP. The evolution towards what is referred to as a *territorial approach* to rural development (Rega, *infra*, Chap. 2) is to be pushed forward with. This represents a change of paradigm and the new policy frame established by the CAP reform offers an opportunity to seize. This rationale encompasses all the key issues recalled in this wrap-up: the need for multidisciplinarity, the consideration of the perceptive dimensions, the importance of trade-offs and of public involvement. It entails the acknowledgement of the complexity of the issues at stake, the conceptualisation of landscape and rural development as spaces of dialectical interactions, and as a consequence the recognition of the impossibility of "neutrality": every decision or intervention on the landscape and the territory is inherently political.

References

Chan KMA, Shaw MR, Cameron DR, Underwood EC, Daily GC (2006) Conservation planning for ecosystem services. Pub Lib Sci Biol 4(11):e379

Domon G (2011) Landscape as resource: consequences, challenges and opportunities for rural development. Landscape Urban Plan 100:338–340

McGranahan DA (2008) Landscape influence on recent rural migration in the US. Landscape Urban Plan 85:228–240

Kinzig AP (2010) Human well being and the natural environment: a focus on ecosystem services. In: Levin SA, Clark WC (2010) (eds) Toward a science of sustainability: report from toward a science of sustainability conference, Airlie Center, Warrenton, Virginia. CID Working Paper No. 196. Center for International Development at Harvard University, 29 Nov 2009—2 Dec 2009, p 105–112

Lefebvre M, Espinosa M et al. (2012) The influence of the common agricultural policy on agricultural landscapes. J. S. a. P. Report, European Commission, Joint Research Center

Talandier M (2009) Mesurer l'impact des aménités naturelles, culturelles et environnementales sur le développement économique local. XLVIe Colloque ASRDLF, Clermont-Ferrand (12/23/2010). http://www.asrdlf2009.com/propositions/final/uploadfinal/talandier.pdf

Turner MG (2010) A landscape perspective on sustainability science. In: Levin SA, Clark WC (2010) (eds) Toward a science of sustainability: report FROM toward a science of sustainability conference, Airlie Center, Warrenton, Virginia. Airlie Center, Warrenton, Virginia. CID Working Paper No. 196. Center for International Development at Harvard University, 29 Nov 2009—2 Dec 2009, p 123–128

Author Biography

The Editor

Carlo Rega holds an MSc in Environmental Engineering and a PhD in Spatial Planning and Local Development. His research interests include spatial/land use planning, agri-environmental policies, ecosystem services, environmental assessment and spatial analysis. From 2008 to 2013, he worked as Post-Doc Fellow and lecturer at the Interuniversity Department of Urban and Regional Studies and Planning of Politecnico di Torino (Italy), where he carried out research on environmental assessment applied to spatial/land use plans and Rural Development Programs. He is currently serving as scientific officer at the Institute for Environment and Sustainability of the European Commission Joint Research Centre, where he works on the assessment of ecosystem services related to land use scenarios, agri-environmental indicators, and biodiversity in agro-ecosystems. He is the author of several papers for peer-reviewed international journals, including *Environmental Impact Assessment Review*, the *Journal of Environmental Planning and Management,* and the *Journal of Environmental Assessment Policy and Management*.

Contributing Authors

Alessandro Bonifazi is an independent researcher and consultant in the field of environmental governance and democracy in spatial policies. He holds a PhD in Urban and Regional Planning from the Polytechnic University of Bari (Italy). He received an MSc in Human Ecology from the *Vrije Universiteit Brussel* (Belgium), after having completed an MSc in Biology at the University of Bari (Italy). He is a member of the *International Association for Impact Assessment*, the *European Evaluation Society*, the Italian *Environmental Analysts Association* and the *International Society for Ecological Economics*. He is the author of several chapters in edited books (published, among others, by Aldershot, Ashgate Publishing Limited and Edward Elgar Publishing), and papers for peer-reviewed international journals, including the *Journal of Environmental Planning and*

C. Rega (ed.), *Landscape Planning and Rural Development*,
SpringerBriefs in Geography, DOI: 10.1007/978-3-319-05759-0,

Management, *Sustainability* and the *Journal of Environmental Assessment Policy and Management*.

Marta Bottero holds a PhD in Geo-Environmental Engineering. She is an Assistant Professor in Project Appraisal and Planning Evaluation at the Interuniversity of Regional and Urban Studies and Planning of Politecnico di Torino (Italy). Her scientific interests are related to sustainability assessment of projects, plans and programmes, with particular reference to the analysis of environmental impacts through specific evaluation techniques. She collaborated to the preparation of the Management Plan of the UNESCO site of "Porto Venere, Cinque Terre and the Islands" where she was involved in the analysis of the economic aspects of landscape. She also participated to the research project "Landscape Indicators" where she studied the use of economic indicators of landscape. Her researches have been published in several scientific journals, such as *Land Use Policy*, *Energy Policy* and *Management of Environmental Quality*.

Chiara Bragagnolo is an environmental scientist with areas of expertise on environmental assessment and planning, GIS and governance of Protected Areas. Her main research interest is the interactions between human and natural systems with the purpose to support decision-making and governance. She obtained her PhD in Environmental Engineering from the University of Trento (Italy). In 2011, she spent a research period at the University of the Azores (Portugal), working on the project Smartparks concerning conservation planning in small islands. She is currently working at the University of Alagoas (Brazil). Her present research challenge is to improve the application of social science to the conservation and the governance of Protected Areas. Her working experience mostly regarded environmental assessment and landscape planning in Northern Italy (including the Province of Trento). She also collaborated to the evaluation of several Regional Operational Programmes supported by European Structural Funds (European Cohesion Policy).

Claudia Cassatella Architect, PhD in Landscape Design, is Assistant Professor in Regional and Urban Planning at Politecnico di Torino (Italy). Her research activities focus on landscape planning, landscape assessment (with particular reference to scenic and perceptive aspects), and the protection of Cultural Heritage, and include applications on regional landscape plans and management plans for World Heritage List and for natural protected areas, commissioned by Public Authorities. With reference to this issue, she teaches in II and III level courses. Currently, she is involved in a European-level coordination of the activities of Landscape Observatories, under the aegis of the European Landscape Network. She is author and editor of several publications, including *Landscape Indicators* (with A. Peano, Springer 2011), *Landscape to be* (Marsilio 2009).

Rinaldo Grittani works at the Agricultural and Environmental Science Department, University of Bari "Aldo Moro" (Italy). He is the head of the photographic section of the newly formed "Laboratory of visual research on landscape". He deals with rural development, environmental protection, agricultural and landscape planning, landscape photography. He holds a PhD in Comparative Agricultural Economics in Eastern Europe from the University Ca' Foscari of Venice (Italy). He received an MSc in Agricultural Sciences from the University of Bari. He is a member of the National Institute of Urban Planning (INU) and of the Territorialist Society (SdT).

Chiara Rizzi international PhD architect, is research fellow and lecturer (since 2011) at the University of Trento. Since 2004 she has been collaborating to researches and projects and giving lectures in national and international seminars. She is author of several chapters in edited books and papers for peer-reviewed international journals. The main theme of her research concerns ecological restoration, mitigation and compensation within a landscape framework. Since 2006 she works as freelance architect and urban planner. Since 2010 she is the president of Architettura Senza Frontiere Abruzzo ONLUS, a non-profit association member of International Committee of ASF, independent network of cooperation and development associations/organizations concerned with the social, cultural and environmental commitment of architecture construction, urbanism, conservation and environmental design in developing countries and poor contexts.

Bianca Maria Seardo holds a PhD in Territorial Planning and Local Development. Her Principal research activity concerns landscape planning, with particular attention to landscape multifunctionality and cultural-ecosystem services issues. After collaborating to the development of the Regional Landscape Plan of Piedmont, her current research deals with rural development and public awareness towards agricultural biodiversity in the context of implementation of Rural Development Programmes.

Stefania Staniscia architect, holds a Master degree in Landscape Architecture (Barcelona School of Architecture—Spain) and an international PhD in Architecture (Venice University IUAV—Italy). She is research fellow and lecturer at the Department of Civil, Environmental and Mechanical Engineering at the University of Trento. Her most important research activities concern spatial and landscape changes due to tourism, decommissioning, and abandonment. The focus is on sensitive and fragile contexts. Islands and mountain areas are her major research fields. Outputs and results of her research activities have been presented in national and international conferences and disseminated through scientific papers and books.

Andrea Tassinari is a freelance Architect—Planner and researcher at Theorema S.a.s. (Venice, Italy). His main research activities concern socio-economic and territorial dynamics and land uses—in terms of consumption and reuse, with special regard to spatial analysis using GIS software and WebGIS based application. In partnership with Theorema S.a.s., he worked for Public Authorities like UNESCO—Office in Venice, Regione Veneto, Regione Puglia, Provincia di Pordenone, and for private companies and research institutes like Cresme Consulting, Fondazione Pellicani, Fondazione Venezia 2000.

Index

C. Rega (ed.), *Landscape Planning and Rural Development*,
SpringerBriefs in Geography, DOI: 10.1007/978-3-319-05759-0,

MIX
Papier aus verantwortungsvollen Quellen
Paper from responsible sources
FSC® C105338

If you have any concerns about our products,
you can contact us on
ProductSafety@springernature.com

In case Publisher is established outside the EU,
the EU authorized representative is:
Springer Nature Customer Service Center GmbH
Europaplatz 3, 69115 Heidelberg, Germany

Printed by Libri Plureos GmbH
in Hamburg, Germany